SAFETY–I AND SAFETY–II

Reviews for
Safety-I and Safety-II

Much more than a technical book. Erik's work is a well documented journey into the multiple interactions between safety, work and human nature. A timely contribution to vindicate human beings and their variability from the one sided focus on the evils of human error. A groundbreaking look at 'the other story' that will certainly contribute to safer and more productive workplaces.

Dr Alejandro Morales, Mutual Seguridad, Chile

Safety needs a new maturity. We can no longer improve by simply doing what we have been doing, even by doing it better. Dr Hollnagel brings forth new distinctions, interpretations, and narratives that will allow safety to progress to new unforeseen levels. Safety–II is more than just incident and accident prevention. A must read for every safety professional.

Tom McDaniel, Global Manager Zero Harm and Human Performance,
Siemens Energy, Inc., USA

Safety–I and Safety–II
The Past and Future of Safety Management

ERIK HOLLNAGEL
University of Southern Denmark

CRC Press
Taylor & Francis Group
Boca Raton London New York

CRC Press is an imprint of the
Taylor & Francis Group, an **informa** business

CRC Press
Taylor & Francis Group
6000 Broken Sound Parkway NW, Suite 300
Boca Raton, FL 33487-2742

© 2014 by Erik Hollnagel
CRC Press is an imprint of Taylor & Francis Group, an Informa business

No claim to original U.S. Government works

Printed on acid-free paper
Version Date: 20160226

International Standard Book Number-13: 978-1-4724-2305-4 (Hardback) 978-1-4724-2308-5 (Paperback)

Visit the Taylor & Francis Web site at
http://www.taylorandfrancis.com

and the CRC Press Web site at
http://www.crcpress.com

Contents

List of Figures

List of Tables

Chapter 1
The Issues

The Need

'Safety' is a word that is used frequently and in many different contexts. Because it is used so often we all recognise it and we all believe that we know what it means – it is immediately meaningful. Because it is immediately meaningful to us, we take for granted that this is the case for others as well. Indeed, when we talk about safety we are rarely, if ever, met with the question 'What do you mean by that?' We therefore make the – unwarranted – inference that other people understand the word 'safety' in the same way that we do. The assumption that we all know and agree on what safety means is so widespread that many documents, standards, guidelines – and even doctoral theses (!) – do not even bother to provide a definition. A search for the etymology of safety, the origin of the word and the way in which its meanings has changed throughout history reveals that it seems to come from the Old French word *sauf*, which in turn comes from the Latin word *salvus*. The meaning of *sauf* is 'uninjured' or 'unharmed', while the meaning of *salvus* is 'uninjured', 'healthy', or 'safe'. (Going even farther back, the roots seem to be in the Latin word *solidus*, meaning 'solid', and the Greek word ηολοσ, meaning 'whole'.) The modern meaning of being safe, as in 'not being exposed to danger', dates from the late fourteenth century; while the use 'safe' as an adjective to characterise actions, as in 'free from risk', is first recorded in the 1580s.

A simple generic definition is that 'safety' means 'the absence of unwanted outcomes such as incidents or accidents', hence a reference to a condition of being safe. A more detailed generic definition could be that *safety is the system property or quality that is necessary and sufficient to ensure that the number of events that could be harmful to workers, the public, or the environment is acceptably low.* Most people will probably agree to this definition without thinking twice about it. A second look, however, makes it clear that the definition is relatively vague because it depends on expressions such as 'harmful to workers' and 'acceptably low'. Yet because each of us finds these expressions meaningful, even though we do interpret them in our own way, we understand something when we encounter them – and naturally assume that others understand them in the same way. The vagueness of the definition therefore rarely leads to situations where the differences in interpretation are recognised.

Is it Safe?

Few who have seen John Schlesinger's 1976 film *Marathon Man* will ever forget the harrowing scene where the villain, Dr Szell, played by Sir Laurence Olivier, tortures the hero, Dustin Hoffman's character Babe, by probing his teeth. While testing existing cavities and even drilling a hole into a healthy tooth, Dr Szell keeps asking the question, 'Is it safe?' While the question is meaningless for Babe, it refers to whether it will be safe for Dr Szell to fetch a batch of stolen diamonds he has deposited in a bank in New York, or whether he risks being robbed by someone – specifically by Babe's brother, who works as an agent for a secret and mysterious US government agency. For Dr Szell, the meaning of 'safe' is whether something will turn out badly when he carries out his plan, specifically whether the diamonds are likely to be stolen. The meaning is therefore the conventional one, namely whether there is a risk that something will go wrong – whether the planned action will fail rather than succeed. But the question could also have been posed differently, namely whether the planned action, the complicated scheme to recover the diamonds, will succeed rather than fail. Here we find the basic juxtaposition between failure and success, where the presence of one precludes the other. But since the absence (negation) of failure is not the same as success, just as the absence (negation) of success is not the same as failure, it does make a difference whether the focus is on one or the other.

On a less dramatic level there are commonly used expressions such as 'have a safe flight' or 'drive safely back' and 'you will be safe here'. The meaning of the first expression is a hope that a journey by plane will take place without any unwanted or unexpected events, that it will be successful and that you will land in Frankfurt – or wherever – as expected. That flying is safe is demonstrated by the fact that on 12 February 2013 it was four years since the last fatal crash in the US, a record unmatched since propeller planes gave way to the jet age more than half a century ago. (It did not last. On 6 July Asiana flight 214 landed short of the runway in San Francisco, killing three passengers and injuring dozens.) The meaning of the second expression, 'drive safely back', is, again, a hope that you will be able to drive home and arrive without any incidents or problems (but not necessarily without having been exposed to any harm). And the meaning of the third expression, 'you will be safe here', is that if you stay here, in my house or home, then nothing bad will happen to you.

What we mean in general by 'being safe' is that the outcome of whatever is being done will be as expected. In other words, that things will go right, that the actions or activities we undertake will meet with success. But strangely enough, that is not how we assess or measure safety. We do not count the tasks where people succeed and the instances when things work. In many cases we have no idea at all about how often something goes right, how often we have completed a flight without incidents, or driven from one place to another without any problems. But we do know, or at least have a good idea of, how many times something has gone wrong, whether it was a delay, the luggage lost, a near miss with another car, a small collision, or other problems. In other words, we know how often we have had an accident (or incident, etc.), but we do not know how often we have not!

The same discrepancy of focus can be found in the management of safety, whether in the deliberate sense of a safety management system or just in the way that we go about what we do in our daily lives. The focus is usually on preventing or avoiding that something goes wrong, rather than on ensuring that something goes right. While it would seem reasonable, if not outright logical, to focus on a positive outcome qua a positive outcome rather than on the absence of a negative outcome, the professional and everyday practice of safety seems to think otherwise. Why this is so is explained in Chapter 3.

The Need for Certainty

One of the main reasons for the dominant interpretation of safety as the absence of harm is that humans, individually and collectively, have a practical need to *be* free from harm as well as a psychological need to *feel* free from harm. We need to *be* free from harm because unexpected adverse outcomes can prevent us from carrying out work as planned and from achieving the intended objectives. Hazards and risks are a hindrance for everyday life and for the stability of society and enterprises, as well as individual undertakings. We need to *be* free from harm in order to survive. But we also need to *feel* free from harm because a constant preoccupation or concern with what might go wrong is psychologically harmful – in addition to the fact that it prevents us from focusing on the activities at hand, whether they be work or leisure. There are many kinds of doubt, uncertainty and worries and, while some of them cannot easily be relieved the doubt about why something has gone wrong can – or at least we presume that is the case. Whenever something happens that we cannot explain, in particular if it was accompanied by unwanted outcomes, we try willingly or unwillingly to find some kind of explanation, preferably 'rational' but if need be 'irrational'. The philosopher Friedrich Wilhelm Nietzsche (1844–1900) described it thus:

> To trace something unfamiliar back to something familiar is at once a relief, a comfort and a satisfaction, while it also produces a feeling of power. The unfamiliar involves danger, anxiety and care – the fundamental instinct is to get rid of these painful circumstances. First principle – any explanation is better than none at all.

There are thus both practical and psychological reasons for focusing on things that have gone wrong or may go wrong. There is a practical need to make sure that our plans and activities are free from failure and breakdowns and to develop the practical means to ensure that. But we also have a psychological need for certainty, to feel that we know what has happened – and also what may happen – and to believe that we can do something about it, that we can master or manage it. Indeed, long before Nietzsche, Ibn Hazm (944–1064), who is considered one of the leading thinkers of the Muslim world, noted that the chief motive of all human actions is the desire to avoid anxiety. This semi-pathological need for certainty creates a preference for clear and simple explanations, expressed in terms that are easy to understand and that we feel comfortable with – which in turn means equally simple methods. It is a natural consequence of these needs that the focus traditionally has been on what I will call the 'negative' side of safety, i.e., on things that go wrong.

Safety as a Dynamic Non-event

One alternative to focusing on unwanted outcomes, which in a very real sense is what safety management does, is, curiously, to focus on what does *not* happen – or rather to focus on what we normally pay no attention to. In an article in *California Management Review* in 1987, professor Karl Weick famously introduced the idea of reliability as a dynamic non-event:

> Reliability is dynamic in the sense that it is an ongoing condition in which problems are momentarily under control due to compensating changes in components. Reliability is invisible in at least two ways. First, people often don't know how many mistakes they could have made but didn't, which means they have at best only a crude idea of what produces reliability and how reliable they are. [...] Reliability is also invisible in the sense that reliable outcomes are constant, which means there is nothing to pay attention to.

This has often been paraphrased to define safety as 'a dynamic non-event', and this paraphrase will be used throughout this book – even though it may be a slight misinterpretation. This is consistent with the understanding of safety as 'the freedom from unacceptable risk' in the sense that a system is safe when nothing untoward happens, when there is nothing that goes wrong. The 'freedom' from the unacceptable risk is precisely the non-event, although it is a little paradoxical to talk about something that is not there. The meaning of 'dynamic' is that the outcome – the non-event – cannot be guaranteed. In other words, we cannot be sure that nothing will happen. It is not a condition of the system that can be established and then left alone without requiring any further attention. Quite the contrary, it is a condition that must constantly be monitored and managed.

Although the definition of safety as a dynamic non-event is very clever, it introduces the small problem of how to count or even notice or detect a non-event. A non-event is by definition something that does not happen or has not happened. Every evening I could, for instance, rightly ask myself how many non-events I have had during the day? How many times was I not injured at work or did not cause harm at work? How many times did I not say or do something wrong or make a mistake? How many cyclists or pedestrians – or cats or dogs – did I not hit when I drove home from work? But I never do, and I guess that no one ever does.

This problem is not just frivolous but actually real and serious. Consider, for instance, an issue such as traffic safety. Every year the traffic safety numbers are provided in terms of how many people were killed in traffic accidents, either directly or as the result of their injuries. And for a number of years, the trend has been that each year the number of dead has been smaller than the year before (see Figure 1.1). Since traffic safety has adopted the goal of zero traffic deaths, that is a development in the right direction. When I read my daily newspaper, I can see how many people were killed in the Danish traffic in the preceding 24-hour period and how many have been killed in the year so far. (Today, 3 August 2013, the total for the preceding 24-hour period is zero, and the total for the year to date is only 94.) But I cannot find out how many people were *not* killed in traffic. Nobody makes a count of that or has the statistics, perhaps because we take for granted that this is the normal outcome and we therefore concentrate on the opposite cases. But knowing how many were not killed is important because it is necessary to know how serious the problem is. We want to ensure that people can drive safely from A to B. We do want to ensure the non-event, but the question is whether it is best done by preventing the 'bad' events or the traffic deaths (which is what we do and which is how we count) or whether it is best done by furthering the 'good' events – which

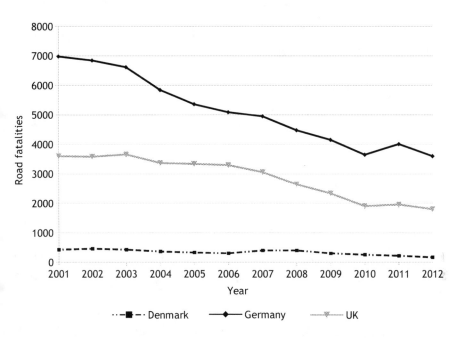

Figure 1.1   Traffic accident statistics for Germany, the UK, and Denmark

is also what we do, but which is not something that we count. Two examples can be used to illustrate these problems.

Signals Passed at Danger

The first example is a train accident that took place in Buizingen, Belgium, on 15 February 2010. Two trains, carrying 250–300 people, collided in snowy conditions during the morning rush hour. The trains apparently collided 'laterally' (i.e., sideways rather than head-on) at a set of points at the exit of Halle station. Eighteen people were killed and 162 injured, with major damage to the tracks as well. The investigation found that one of the trains had passed a red signal without stopping (a situation that happens so often that it has acquired its own name: SPAD or Signal Passed At Danger), and that this could be a contributing cause to the collision, although not the only one.

Further investigation revealed that there were 130 SPAD events in Belgium in 2012, of which one third were serious. (In 2005 there were 68 SPADs, but the number had been increasing each year.) But it was also estimated that there were about 13,000,000 cases per year of trains stopping at a red signal, which means that the probability of a SPAD is 10^{-5}. A value of 10^{-5} means that the system is safe, although not ultrasafe. However, for activities where humans are involved, 10^{-5} is not unacceptable. (The probability of an accident calculated from the same numbers is 7.7×10^{-8}, which is probably as good as it gets.) In this case it was possible to find, or rather to estimate, how many times the activity succeeded, and thereby get an idea about how serious the event was – not in terms of its outcomes, which were serious and even tragic, but in terms of its occurrence.

In relation to the accident at Buizingen, the fact that 13,000,000 trains stopped at a red signal does not mean that they all stopped in the same way. An elevator is a purely mechanical system that will stop in the same way whenever it reaches the floor it is going to. (Although even here there may be variability due to load, wear and tear, adjustments, maintenance, or other conditions.) But a train is a human–machine system, which means that it is stopped by the train engineer rather than by a mechanism. The way in which a train stops is therefore variable, and it is important to know the variability in order to understand how it is done – and how it can fail. By analogy, try to look at how a driver brakes when the car comes to a red light (but do it from the pavement, not while driving yourself). The way in which a car brakes depends on the load, the driver, the weather, the traffic conditions, etc. It can be, and is, done in many different ways, and all of them usually achieve their goal.

Switching from Left to Right

An interesting, although rather unique, case of a situation where the number of non-events was known with certainty was 'Dagen H' in Sweden. On this day, Sunday, 3 September 1967, Sweden changed from driving on the left-hand side to driving on the right-hand side. In order to accomplish that, all non-essential traffic was banned from the roads from 01:00 to 06:00. Any vehicle on the road during that time, for instance the fire brigade, ambulances, the police and other official vehicles, had to follow special rules.

All vehicles had to stop completely at 04:50, then carefully change to the other side of the road and remain there until 05:00, when they were allowed to proceed. (In the major cities, the driving ban was considerably longer in order for working crews to have sufficient time to reconfigure intersections.)

Because there was no traffic, or only so little traffic that it in principle could be monitored and accounted for, it can be said with certainty that there were no non-events during the change. Or at least the number of non-events was countable, had anyone bothered to count them. And since there were no non-events, since cars were not allowed to drive, there could not be any events either, i.e., no collisions between cars. (In absolute terms this lasted only for the ten minutes from 04:50 to 05:00, but in practice it lasted for the five hours between 01:00 and 06:00.)

Even when people agree that safety is a dynamic non-event, the practice of safety management is to count the events, i.e., the number of accidents, incidents, and so forth. By doing that we know how many events there have been, but not how many non-events. We may, however, easily turn the tables, by defining safety as a dynamic *event*. The event is now that an activity succeeds or goes well (that we come home safely, that the plane lands on time, etc.), and we are obviously safe when that happens. The non-event consequently becomes the situation when this does *not* happen, i.e., when things go wrong. We can count the non-events, i.e., the non-successes or failures, just as we have usually done. But we can now also count the events, the number of things that go right, at least if we make the effort.

The Measurement Problem

In order to know that we are safe – not just subjectively or psychologically, but also objectively or practically – industry and society need some way of demonstrating the presence of safety. In practice this means that there must be some way of quantifying safety.

Strictly speaking, it must be possible to confirm the presence of safety by means of *intersubjective verification*. To the extent that safety is an external, public phenomenon, the way in which it is experienced and described by one individual must correspond to or be congruous with how it is experienced and described by other individuals. In other words, it must be possible for different individuals to talk about safety in such a manner that they can confirm that they understand it in the same way. It must specifically be possible for an individual carefully to describe safety (in words or otherwise) so that others can confirm or verify that their experiences of the phenomenon, their understanding of safety, 'fits' the description. It is not just a question of whether people recognise the term 'safety' and subjectively experience that they know what it means. Intersubjective verification means going beyond the lack of disagreement ('I don't know what this means') to an explicit act of communication in order to establish that the term is not just recognised but that it actually means the same to two or more people.

There is a genuine practical need to *avoid* that things go wrong. There is also a genuine practical need to *understand* why things go wrong, why systems fail, why people are hurt, why property or money are lost, etc. And there is finally a genuine practical need to *measure* how safe we are or how safe a certain kind of activity is. This is why we incessantly – and sometimes indiscriminately – collect data and look for assurance in the safety statistics. Because 'No man is an Iland, intire of it selfe' (John Donne), we need to know how safe something 'really' is in order to decide how we should act in relation to it, often as a question of whether we should do it or not.

A famous example of general safety statistics is a list compiled by Bernard Cohen, Professor Emeritus of Physics at the University of Pittsburg, parts of which are shown in Table 1.1. The list provides a rank order of various activities that correspond to the same risk, namely an increased probability of death by 10^{-6}. (The number itself is not very meaningful, not even to a specialist.) The list shows that it is just as dangerous to go for a 10-mile ride on a bicycle as it is to travel 1,000 miles by plane, or smoke 1.4 cigarettes – not per day but in total!

Although the items on the list represent the same level of numerical or objective risk, they do not represent the same level of subjective risk. If that was the case, then a person who is willing to go for a 10-mile bicycle ride should be equally willing to undertake any of the other activities, for instance eat 100

Table 1.1   Comparison of different activities with the same risk

Spend three hours in a coal mine (risk of having an accident).	Travel 10 miles by bicycle (risk of having an accident).
Travel 300 miles car (risk of having an accident).	Travel 1,000 miles by jet air-plane (risk of having an accident).
Smoke 1.4 cigarettes.	Live two months with a smoker.
Eat 100 charcoal-broiled steaks.	Drink 30 cans saccharine soda.
Live 20 years near PVC plant (cancer from vinyl chloride).	Live 150 years at 20 miles from a nuclear power plant.
Live two months in Denver (cancer from high average radiation).	Live five miles from nuclear plant for 50 years (nuclear accident).

charcoal-broiled steaks or live five miles from a nuclear power plant. Few would, however, choose to do so, one reason being that it is difficult to contemplate a risk in isolation. Other reasons have to do with individual values, morals, biases, personality type, etc.

The Regulator Paradox

But quantifying safety by measuring what goes wrong will inevitably lead to a paradoxical situation. The paradox is that the safer something (an activity or a system) is, the less there will be to measure. In the end, when the system is perfectly safe – assuming that this is either meaningful or possible – there will be nothing to measure. In control theory, this situation is known as the 'regulator paradox'.

In plain terms the fundamental regulator paradox means that if something happens rarely or never, then it is impossible to know how well it works. We may, for instance, in a literal or metaphorical sense, be on the right track but also be precariously close to the limits. Yet if there is no indication of how close, it is impossible to improve performance. The fundamental regulator paradox has been described as follows:

> The task of a regulator is to eliminate variation, but this variation is the ultimate source of information about the quality of its work. Therefore, the better the job a regulator does the less information it gets about how to improve.

While pragmatically it seems very reasonable that the number of accidents should be reduced as far as possible, the regulator paradox shows that such a goal actually is counterproductive in the sense that it makes it increasingly difficult to manage safety. Safety management, like any other kind of management, is fundamentally an issue of regulation (or control). (The origin of 'manage' is the Latin word *manus*, or hand, the meaning being to have strength or have power over something.) The essence of regulation is that the regulator makes an intervention in order to steer or direct the process in a certain direction. (The simplest example is a vehicle, such as an automobile, being driven by a driver who is in direct control of speed and direction – although gradually less and less.) But if there is no response to the intervention, if there is no feedback from the process, then we have no way of knowing whether the intervention had the intended effect. (It can also be critical for safety management if the feedback is slow or delayed, since that makes it difficult to determine whether or not the consequences of an intervention have occurred.) A further problem in relation to safety is that a lack of indications is often used as an excuse for redistributing scarce resources away from safety to some other area. In other words, a high level of safety (as defined by a low count of things that go wrong) is often used to justify a reduction in safety efforts.

An example of the latter kind of reasoning is provided by a statement released by the Danish Energy Agency (Energistyrelsen) in the autumn of 2012. The DEA noted that the compulsory annual energy measurement of 25,000 boilers in private homes corresponding to 10 per cent of the boilers in Denmark, had found only 10 cases in which further overhaul was required. The DEA expressed surprise by this finding and stated that it would consider whether the requirement, which had its source in a European Union (EU) Directive, should be amended. The DEA calculated that the 10 cases would correspond to a total of around 100 boilers in Denmark, and that it would look critically at whether it would be too great an effort to examine all 250,000 boilers. It was felt that the system of inspections might be 'overshooting the mark', hence that the cost was unnecessary.

An even clearer example is provided by the September 2012 decision of the UK's coalition government to confine health and safety inspections to high-risk sites only. This brought to an end the automatic inspection of both low and medium-risk industries and resulted in 11,000 fewer inspections each year. It meant a reduction in the regulatory burden on business and also contributed to the planned 35 per cent cut in the budget of the Health and Safety Executive (HSE). The argument was that such inspections were unnecessary for low-risk businesses. The British Chambers of Commerce also welcomed the decision, and Adam Marshall, Director of Policy and External Affairs, said:

> Ensuring that low-risk workplaces are exempted from inspections is a sensible change that will save employers time and money without reducing the safety of workers. [...] [We have] long argued for a risk-based approach to health and safety with industry-specific rules and a light-touch regime for those operating a low-risk workplace.

Some, including the trade unions, were not convinced that safety really was so high that the inspections were no longer needed. The Trades Union Congress (TUC) General Secretary Brendan Barber said:

> Health and safety regulation is not a burden on business; it is a basic protection for workers. Cutting back on regulation and inspections will lead to more injuries and deaths as result of poor safety at work [...] Some of the 'low risk' workplaces identified by the government, such as shops, actually experience high levels of workplace injuries. This will only get worse if employers find it easier to ignore safety risks.

In both cases, the consequence is that a source of information is eliminated, which means that the possibility of regulating or managing is reduced. It is also a little ironic that one argument for reducing the number of inspections is that this means the adoption of a risk-based approach, unless, of course, risk is used as a euphemism for ignorance.

European Technology Platform on Industrial Safety

The European Technology Platform on Industrial Safety – which uses the slightly unfortunate acronym ETPIS – is another good illustration of how increased safety is defined by a diminishing number of accidents. The premise for the Technical Platform, launched in 2004, was the recognition that safety was a key factor for successful business (in Europe, but presumably also elsewhere) as well as an inherent element of business performance, and that it would therefore be necessary to improve safety in order to improve business performance. It was proposed to do this by introducing a new safety paradigm in European industry. According to the Technical Platform, improved safety performance would show itself in terms of reduction of *reportable accidents at work, occupational diseases, environmental incidents and accident-related production losses* – in other words, in terms of a diminishing number of adverse outcomes.

The aim of the Technical Platform is to have structured self-regulated safety programmes in all major industry sectors in all European Countries by 2020, leading to 'incident elimination' and 'learning from failures' cultures – also called an accident-free mindset. The concrete targets, as set in 2004, were a 25 per cent reduction in the number of accidents by 2020, followed by a continued annual accident reduction at a rate of 5 per cent or better.

The Technical Platform is typical by having a reduction in the number of accidents as its main objective, and by defining this objective in a quantifiable manner. In other words, by setting concrete and quantifiable targets. There is, of course, nothing objectionable in wanting to reduce the number of accidents, but it may be discussed whether the chosen approach is the most appropriate one – quite apart from the fact that it seems to rely on intangible qualities such as an 'incident elimination' culture. However, as is argued above, by having the reduction of accidents, and thereby the elimination of feedback, as the main objective, the Technical Platform will – if successful – effectively remove the very means by which it can be managed, hence make it more difficult to achieve its stated goal. In this context it is only a small consolation that it may take years before it happens – if it happens at all. (Another example, and just one among many, is the Single European Sky ATM Research or SESAR programme for building the future European air traffic management system, where one of the four targets is to improve safety by a factor of 10. In practice this *improvement* will be measured by a *reduction* in the number of reported accidents and incidents.)

Being Risky or Being Safe: The Numbers Game

While it is clear that a safety management system requires some feedback in order to be able to do its work, it does not follow that the feedback must refer to adverse outcomes, or that it must be quantitative. The convenience of measurements is that they make it easy to compare different activities or performances; and that they also make it possible to follow the same activity over time to see whether it has become better or worse. (Measurements can thus be seen as representing an efficiency–thoroughness trade-off. Instead of directly observing, analysing, and interpreting how a set of activities are carried out – which is thorough, but not very efficient – a limited set of indicators or signs can be recognised and tallied – which is efficient but not very thorough.)

When looking for something to measure, counting discrete events is an attractive option, not least if they are conspicuous in some way. It seems obvious that we can count various types of events, such as accidents, incidents, near misses, etc. – provided that they can be defined in an operational sense. So if the number of things that go wrong are taken to represent safety – or rather, the absence of safety – then safety can be measured. Problems arise, however, when we think of safety as a 'dynamic non-event'. One reason is that we do not have a distinct typology or taxonomy of non-events at our disposal. Another is that it is impossible to observe, let alone count, something that does not happen.

While the requirement of a clear and operational definition of the types of events to count may seem an easy problem to solve, it is in fact rather difficult as the following examples show.

1. Consider, for instance, the risk of being kidnapped. If the risk is high, the situation is clearly unsafe, and vice versa. For kidnapping, official figures from the United Nations show that there were 17 kidnaps per 100,000 people in Australia in 2010 and 12.7 in Canada, compared with only 0.6 in Colombia and 1.1 in Mexico. For most people that will come as a surprise. Can it really be the case that the risk of being

kidnapped is 28 times higher in Canada than in Colombia? The reason for this dramatic difference is to be found in how a kidnapping case is defined, with the definitions being vastly different between different countries. In Canada and Australia parental disputes over child custody are included in the figures. If one parent takes a child for the weekend, and the other parent objects and calls the police, the incident will be recorded as a kidnapping. The two countries would therefore be a lot 'safer' if the numbers only included 'real' kidnappings, as they presumably do in Mexico and Colombia.

2. Other statistics show that Sweden has the highest rape rate in Europe. In 2010, the Swedish police recorded the highest number of offences – about 63 per 100,000 inhabitants – of any force in Europe, and the second highest in the world. The number was twice the rate in the US and the UK, three times higher than the number of cases in neighbouring Norway, and 30 times the number in India. Can we therefore conclude that Sweden is a much more dangerous place for women than these other countries? No we cannot, because the numbers are not comparable. Police procedures and legal definitions vary widely among countries, and the case numbers represent the local culture rather than an objective count. The reason for the high numbers in Sweden is that every case of sexual violence is recorded separately. If a woman comes to the police and complains that her husband or fiancé has raped her almost every day during the last year, each of these cases is recorded, resulting in more than 300 events, while many other countries would just record the woman's report as one event.

3. The situation is not much better for railway accidents. In Denmark, the national railway company (DSB) briefly defines accidents as events that involve serious injuries or significant damage to equipment or infrastructure, in practice meaning damage amounting to millions of Danish Crowns. In the US, train accidents and incidents include all events reportable to the US Department of Transportation, Federal Railroad Administration under applicable regulations. This comprises collisions, derailments and other events involving

the operation of on-track equipment and causing reportable damage above an established threshold ($6,700 in 2003); highway-rail grade crossing incidents involving impact between railroad on-track equipment and highway users at crossings; and other incidents involving all other reportable incidents or exposures that cause a fatality or injury to any person, or an occupational illness to a railroad employee. In Japan, Japan Railways defines a 'big accident' as one that either involves a loss of more than 500,000 Japanese yen, corresponding to 30,000 Danish Crowns) or causes a delay of more than 10 minutes to the first Shinkansen of the day. (Few other countries would bother to classify a 10-minute train delay as an accident.) Finally, the Accident Manual of India's West Central Railway defines a serious accident as an 'accident to a train carrying passengers which is attended with loss of life or with grievous hurt to a passenger or passengers in the train, or with serious damage to railway property of the value exceeding Rs. 2,500,000' – but with a number of interesting exceptions, such as 'Cases of trespassers run over and injured or killed through their own carelessness or of passengers injured or killed through their own carelessness'. It would clearly be difficult to compare railway safety in Denmark, the US, Japan, and India simply by looking at the number of accidents!

So Where Does This Leave Us?

Human beings have an undisputed need to be and to feel safe – individually and collectively. This introductory chapter has argued that it is important to be thorough in trying to find solutions to that need. The immediate – and 'obvious' – solution is to focus on what goes wrong, to eliminate that, and to verify the outcome by counting and comparing the number of accidents before and after an intervention has been made. While this may be efficient in the sense that it quickly – but alas only temporarily – reduces the feeling of uncertainty, it is not a thorough solution in the long run. Had it been, we would no longer be struggling with the same problems, or indeed an increasing number of ever more diverse problems. (Neither would it have been necessary to write this book.) The right solution therefore lies elsewhere, and the following chapters show the way step by step.

Comments on Chapter 1

The quote from Friedrich Wilhelm Nietzsche (1844–1900) is from *The Twilight of the Idols*, which was written in 1888 and published in 1889. This book also contains a description of the 'Four Great Errors', which include confusing the effect with the cause. Although aimed at philosophy (and philosophers), it is not irrelevant for safety management.

The famous definition of 'reliability as a dynamic non-event' is from Weick, K.E. (1987), Organizational culture as a source of high reliability, *California Management Review*, 29(2), 112–27. As is explained above, it is often quoted as 'safety is a dynamic non-event'.

Bernard Cohen (1924–2012) became interested in how to explain risks in connection with his study of the effects of low-level radiation. As part of that he published a 'Cataloge of Risk', which has been widely used and was revised several times. The latest version is published in *The Journal of American Physicians and Surgeons*, 8(2), 50–53.

A detailed account of the rail accident in Buizingen, Belgium, can be found in the investigation report (in French) from *Organisme d'Enquête pour les Accidents et Incidents Ferroviaires*, published in 2012. The URL is http://www.mobilit.belgium.be/fr/traficferroviaire/organisme_enquete/.

The regulator paradox has been described in many places, for instance in Weinberg, G.M. and Weinberg, D. (1979), *On the Design of Stable Systems*, New York: Wiley. It also appears in several places on the web, including http://secretsofconsulting.blogspot.com/2010/10/fundamental-regulator-paradox.html.

Extensive information about The European Technology Platform on Industrial Safety is available from http://www.industrialsafety-tp.org/. Despite being more than halfway through its programme, the project still subscribes to the goals and strategies set out in 2004, at least according to the website.

The section headed 'Being Risky or Being Safe: The Numbers Game' refers to the idea of the efficiency–thoroughness trade-off. This has been presented in detail in Hollnagel, E. (2009), *The ETTO Principle: Efficiency–thoroughness Trade-off: Why Things That Go Right Sometimes Go Wrong*, Farnham: Ashgate. *The ETTO Principle* introduces the idea of performance variability, which is a central topic in this book, and references to this trade-off will be made several times in the following chapters.

Chapter 2
The Pedigree

The History

The development of thinking about safety can be described in terms of the development in the thinking about causes and of the development in the thinking about 'mechanisms'. The 'causes' refer to the socially accepted origins or reasons for accidents, specifically the elements, parts, or components that can fail and how they fail. The thinking about causes is thus closely linked to the notion of causality and to failure, both of which will be considered in detail later. The 'mechanisms' refer to the ways in which a cause, or a combination of causes, can lead from the initial failure or malfunction to the consequence. The causes account for *why* accidents happen and the 'mechanisms' for *how* accidents happen. While the two clearly are not independent, the developments in thinking about causes have not been synchronised with the developments in thinking about 'mechanisms'. The set of possible causes reflect the changes in the technology we use as well as the changes to the nature of the systems we employ – from 'purely' technical systems, such as a steam engine, via socio-technical systems such as a train dispatch centre, an oil rig, or a stock exchange, to socio-technical habitats such as an urban transportation system, energy provision, or the financial market.

The change in the nature of causes in a sense reflects the changes in what the constituents of systems of work are, what they are made of, and the reliability of the various constituents. When there was no electricity, there could be no short circuits; when there was no radio communication, there could be no transmission errors; when there was no software, there could be no software errors, etc. From the very beginning, systems have had two constituents, namely people and technology of some kind. (Both are, of course, still present, efforts to introduce complete automation notwithstanding.) Of these, technology has until recently been the primary suspect when something went wrong. For a long time – for millennia in fact – technology was seen as the most important system component, and technology could fail. Technical components were therefore seen as the primary cause – as the usual suspect. But as we moved into the nineteenth and twentieth centuries two things happened. One was that technology became more and more reliable, hence less and less likely to fail. The second was that other constituents became important, first various forms of energy; and later, as work changed from being manual to being mechanised, various forms of control systems and self-regulation. (Nominally these could still be seen as technology, but in relation to safety it was often the logic rather than the technology that was the problem.) As time went by, the reliability of technical components steadily improved until they became so reliable that they moved from being a primary to becoming a secondary – or even lower – concern. The thinking about causes is thus in a way a product of the state of the art, in the sense that the least reliable component (group) is the most likely among causes. (It also happens to be the type of constituent that we know the least about, which may be a reason for the unsatisfactory reliability.) In the thinking about types of causes, we see a development that goes from technology to the human factor and, most recently, to the organisations and to culture. This can also be seen as a development from the concrete to the intangible or incorporeal.

The change in 'mechanisms', i.e., in the preferred explanations of how unwanted outcomes can occur, has only been driven indirectly by scientific and technological developments. The changes so far seem to have happened when the current way of thinking – the current paradigm, if one dare use so grand a term – has reached a dead end. This means that something has happened that could not be explained in a satisfactory manner by the currently accepted approaches. (It might be going too far to call this a paradigm shift, although there are similarities to how the American physicist Thomas Kuhn, who coined this term, used it.) In the thinking about how accidents happen, we similarly see a development from simple, individual causes and a single linear progression through a chain of causes and effects, to composite causes where several lines of progression may combine in various ways and, finally, to the non-linear types of explanation which so far represent the most advanced stages. At each stage of this development, the new ways of understanding and explaining accidents were intended to supplement rather than to replace what was already there.

Through the ages, the starting point for safety concerns has been the occurrence, potential or actual, of some kind of adverse outcome, whether it has been categorised as a risk, a hazard, a near miss, an incident, or an accident. Historically speaking, new types of accidents have been accounted for by introducing new types of causes (for example, metal fatigue, 'human error', software failure, organisational failure) rather than by challenging or changing the basic underlying assumption of causality. Since humans have a partiality towards simple explanations, there has also been a tendency to rely on a single type of cause. The development has therefore been to replace one dominant type by another, rather than to combine them or consider them together. The advantage of sticking with a single type of cause is that it eliminates the need to describe and consider the possible interactions among causes or their dependence on each other. In consequence of this we have through centuries become so accustomed to explaining accidents in terms of cause–effect relationships – simple or compound – that we no longer notice it. And we cling tenaciously to this tradition, although it has becomes increasingly difficult to reconcile with reality.

It is only recently that we have found it necessary to change not only the possible causes but also the way they create their effects – in other words the 'mechanisms' that we use to explain how and why things happen. This unfortunately means that we have to give up the notion of causality in the conventional sense. Things do still happen for a reason, of course, but that reason is no longer a simple occurrence such as the malfunctioning of a component – and not even a combination of the malfunctioning of multiple components. The reason can instead be a condition or situation that only existed for a brief moment of time, yet long enough to affect some of the actions or activities that followed.

The Three Ages in Safety Thinking

The realisation that things can go wrong is as old as civilisation itself. The first written evidence is probably found in the Code of Hammurabi, produced circa 1760 BC, which even includes the notion of insurance against risk ('bottomry') by which the master of a ship could borrow money upon the *bottom* or keel of it, so as to forfeit the ship itself to the creditor, if the money with interest was not paid at the time appointed at the ship's safe return.

It was nevertheless not until after the second industrial revolution in the late eighteenth century that risk and safety became a concern not just for the people who did the work, but also for those who designed, managed and owned it. (The first industrial revolution was the transition from a hunting and gathering lifestyle to one of agriculture and settlement, which started around 12,000 years ago.) Professors emeriti Andrew Hale and Jan Hovden have described the development of safety by distinguishing between three ages, which they named 'the age of technology', 'the age of human factors', and 'the age of safety management' (Figure 2.1).

The First Age

In the first age, the dominant threats to safety came from the technology that was used, both in the sense that the technology (mainly steam engines) itself was clunky and unreliable, and in the sense that people had not learned how systematically to analyse and guard against the risks. The main concern was to find the technical means to safeguard machinery, to stop explosions and to prevent structures from collapsing. Although the concern for accidents is undoubtedly as old as human civilisation itself, it is generally agreed that the beginning of the industrial revolution (usually dated to 1769) introduced new risks and gradually also a new understanding of risks.

One of the earliest examples of a collective concern for safety and risk was the US *Railroad Safety Appliance Act* of 1893, which argued for the need to combine safety technology and government policy control. The railway was one of the first conditions in which 'innocent bystanders' – meaning train passengers – were

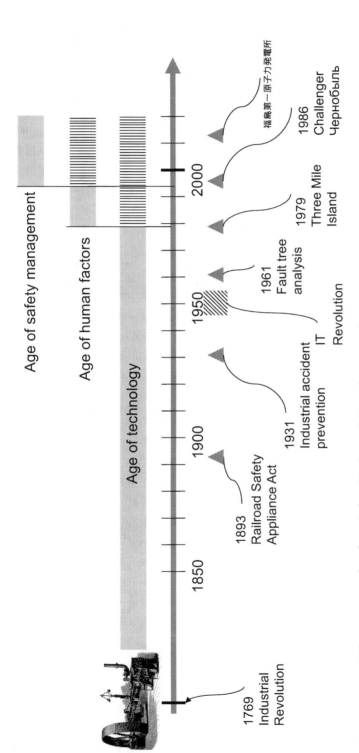

Figure 2.1   Three ages of safety (after Hale and Hovden, 1998)

exposed to industrial risks other than in their line of work. (As a tragic illustration of that, a passenger – Mr William Huskisson, Member of Parliament for Liverpool, a former Treasurer of the Navy, President of the Board of Trade, and Cabinet Minister – was run over by a train and killed on the opening of the Liverpool and Manchester railway on 15 September 1830. This earned him the unenviable distinction of being the world's first railway casualty.) A railway is also an excellent early example of an organised activity that includes all the facets that characterise today's organisational environment, including the needs to plan activities and operations, to train staff, to procure materials and tools, to maintain equipment, to coordinate activities vertically and horizontally, to develop specialised functions and entities, to monitor and control daily operations, etc. In other words, a socio-technical system.

Perhaps the most prominent example of a concern for safety was Heinrich's highly influential book, *Industrial Accident Prevention* of 1931. Yet despite the need for reliable equipment that exists in every industry, the need for reliability analysis only became widely recognised towards the end of the Second World War. One reason was that the problems of maintenance, repair and field failures of the military equipment used during the Second World War had become so severe that it was necessary to do something about them. Another reason was that new scientific and technological developments made it possible to build larger and more complicated technical systems – in depth and in breadth – that included extensive automation. Prime among these developments were digital computers, control theory, information theory, and the inventions of the transistor and the integrated circuit. The opportunities for improved productivity offered by these developments were eagerly awaited and quickly adopted by a society obsessed with 'faster, better, cheaper' – although that particular phrase did not come into use until the 1990s. The systems that resulted were, however, often so difficult to understand that they challenged the human ability both to account for what went on inside them and to manage it.

In the civilian domain, the fields of communication and transportation were the first to witness the rapid growth in scope and performance as equipment manufacturers adapted advances in electronics and control systems (these developments are described further in Chapter 6). In the military domain, the development of missile defence systems during the Cold War, as well as the beginning of the space programme, relied on equally complicated technological systems. This created a need for proven methods by which risk and safety issues could be addressed. Fault Tree analysis, for instance, was originally developed in 1961 to evaluate the Minuteman Launch Control System for the possibility of an unauthorised missile launch. A Fault Tree is a formal description of the set of events that in combination may lead to a specific undesired state, which in the Fault Tree is called the 'top' event. The Fault Tree method provides a systematic way of analysing how a specific undesired outcome might happen, in order to develop precautions against it.

A generic fault tree that describes how an accident – details unspecified – can happen is known as 'the anatomy of an accident', see Figure 2.2. According to this description an accident begins when an unexpected event occurs while the system is working normally. The unexpected event can itself be due to an external event or a latent condition that for some reason suddenly becomes manifest. Unless the unexpected event can be immediately neutralised, it will move the system from a normal to an abnormal state. In the abnormal state attempts will be made

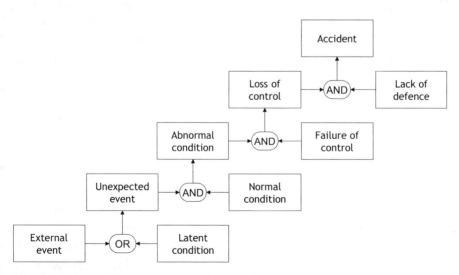

Figure 2.2   The anatomy of an accident

to control the failure. If this control fails, the system will enter a state of loss of control, which means that there will be some unwanted occurrences. Usually, even this possibility has been anticipated and specific barriers have been provided. It is only when these barriers are missing or fail to work that the adverse consequences occur, which means that the accident happens.

Other methods such as Failure Mode and Effects Analyses (FMEA) and Hazard and Operability Analysis (HAZOP) were developed not just to analyse possible causes of hazards (and, later on, causes of accidents), but also systematically to identify hazards and risks before a system was taken into operation or when a major change was considered.

By the late 1940s and early 1950s, reliability engineering had become established as a new and separate engineering field, which combined the powerful techniques of probability theory with reliability theory. This combination became known as probabilistic risk assessment (PRA), sometimes also called probabilistic safety assessment (PSA). PRA was successfully applied to the field of nuclear power generation where the WASH-1400 'Reactor Safety Study' became the defining benchmark. This study considered the course of events which might arise during a serious accident at a large modern Light Water Reactor, using a fault tree/event tree approach. The WASH-1400 study established PRA as the standard approach in the safety assessment of modern nuclear power plants, from where the practice gradually spread to other industries with similar safety concerns. The focus of PRA was, however, on the technology rather than on humans or the organisation.

The Second Age

The feeling of having mastered the sources of risks so that the safety of industrial systems could be effectively managed was rather abruptly shattered by the disaster at the Three Mile Island (TMI) nuclear power plant on 28 March 1979. Before this happened, the consensus had been that the use of established methods such as HAZOP, FMEA, Fault Trees and Event Trees would be sufficient to ensure the safety of nuclear installations. The nuclear power plant at TMI had itself been through a PRA and had been approved by the US Nuclear Regulatory Commission as safe. After this disaster it became painfully clear that something was missing in this approach, namely the human factor. The human factor had been considered in human–machine system design and operation since human factors engineering had started in the US as a speciality of industrial psychology in the mid-1940s. (In the US the Human Factors and Ergonomics Society was founded in 1957. In Europe, the history of human factors is a bit longer, as is shown by the journal *Le Travail humain*, which was started already in 1937, and the UK Ergonomics Research Society, which was founded in 1946.) Although the experience of the US Army during the Second World War had shown clearly that so-called 'pilot errors' could be greatly reduced by paying attention to the design of displays and controls, human factors was not seen as being crucial for safety by the industry in general. Instead, human factors engineering focused mainly on the efficiency or productivity side of system design. After the scientific and technological breakthroughs in the 1940s that vastly increased the capability of technology, humans came to be seen as too imprecise, variable, and slow a match for the technology, and hence as a limitation on system productivity. The generic solutions were design, training and automation, of which the latter in particular led to a dependence on technological fixes that in the end became self-defeating. Although technological and engineering innovations during the 1960s and 1970s made technology both more powerful and more reliable, the accidents continued to increase, in terms both of number and magnitude, reaching a pinnacle with what happened at TMI. In the general view, humans came to be seen as failure-prone and unreliable and so as a weak link in system safety. The 'obvious' solution was to reduce the role of humans by replacing them by automation, or to limit the variability of human performance by requiring strict compliance.

Since PRA by that time had become established as the industry standard for how to deal with the safety and reliability of technical systems, it was also the natural starting point when the human factor needed to be addressed. Extending PRA to include human factors concerns led to the development of a number of methods for Human Reliability Assessment (HRA). At first, the existing methods were extended to consider 'human errors' in the same way as technical failures and malfunctions, but these extensions were soon replaced by the development of more specialised approaches. The details of this development have been described extensively in the professional literature, but the essence is that human reliability became accepted as a necessary complement to system reliability – or rather that technology-based thinking about reliability engineering was extended to cover the technological and the human factors. The use of HRA quickly became established as the standard analysis for nuclear power plant safety, but despite many efforts there have never been any fully standardised methods – or even a reasonable agreement among the results produced by different methods. The idea that 'human error' could be used to explain the occurrence of adverse events was eagerly adopted by other industries, and the development of models and methods quickly took on a life of its own.

The development of technical risk analysis also led to a gradual intellectual separation between reactive safety (accident investigation) and proactive safety (risk assessment). For the latter it was taken for granted that risks were a question of probabilities, of the likelihood or otherwise of something happening. The focus was therefore on the probability of future events, particularly on how likely it was that a specific failure or malfunction would occur. For accident investigation, probability was not an issue. When something had happened, then something had happened. The primary concern was therefore to establish the cause – or causes – and the focus was on causality. Since causes were supposed to be definitive rather than probable, it was considered unsatisfactory if one could only say that something possibly could have been the cause.

The Third Age

Whereas belief in the efficacy of technological failures as causes went unchallenged for more than 200 years, a similar belief in the human factor lasted barely a decade. There are two main reasons for this. There was first of all a growing doubt of the idea that health and safety could be ensured by a normative approach, for instance simply by matching the individual to the technology (as in classic human factors engineering and Human–Machine Interaction design). Secondly, several accidents had made it painfully clear that the established approaches, including PRA-HRA and numerous 'human error' methods, had their limitations. Although a revision of the established beliefs was less dramatic than the transition from the first to the second age, accidents such as the Space Shuttle *Challenger* Disaster and the explosion of reactor number four at the Chernobyl nuclear power plant, which both happened in 1986, and in retrospect also the 1977 collision of two Boeing 747 airliners on the runway at Tenerife North Airport, made it clear that the organisation had to be considered over and above the human factor. One consequence was that safety management systems have become a focus for development and research, and even lend their name to the third age: 'the age of safety management'.

The attempts to extend the established basis for thinking about risk and safety, i.e., reliability engineering and PRA to cover also organisational issues were, however, even less straightforward than the attempts to include the human factor in the linear causality paradigm. Whereas the human in some sense could be seen as a machine, a tradition that goes back at least to the French physician and philosopher Julien Offray de la Mettrie (1709–1751) and which was given new life by the popular analogy between the human mind and a computer, the same was not the case for an organisation. It was initially hoped that the impact of organisational factors could be determined by accounting for the dependencies that these factors introduced among probabilistic safety assessment parameters, using Performance Shaping Factors as an analogy. But after a while it became clear that other ways of thinking were required. The school of High Reliability Organisations (HRO) argued that it was necessary to understand the organisational processes required to operate tightly coupled technological organisations with non-linear functioning. Other researchers pointed out that organisational culture had a significant impact on the possibilities for organisational safety and learning, and that limits to safety might come from political processes as much as from technology and human factors.

At present, the practices of risk assessment and safety management still find themselves in the transition from the second to the third age. On the one hand it is realised by many, although not yet by all, that risk assessment and safety management must consider the organisation either as specific organisational factors, as safety culture, as 'blunt end' factors, etc.; and furthermore, that when accidents are attributed to organisational factors, then any proposed interventions to change these factors must also be the subject of a risk assessment, since no intervention can be 'value neutral'. On the other hand it is still widely assumed that the established approaches as practised by engineering risk analysis either can be adopted directly or somehow can be extended to include organisational factors and organisational issues. In other words, organisational 'accidents' and organisational failures are today seen as analogous to technical failures, just as human failures were in the aftermath of the TMI disaster. And since HRA had 'proved' that the human factor could be addressed by a relatively simple extension of existing approaches, it seemed reasonable to assume that the same was the case for organisational factors. This optimism was, however, based on hopes rather than facts and in the end turned out to be completely unwarranted. It is becoming increasingly clear that neither human factors nor organisational factors can be adequately addressed by using methods that follow the principles developed to deal with technical problems, and that it is a gross oversimplification to treat them as 'factors'. There is therefore a need to revise or even abandon the commonly held assumptions and instead take a fresh look at what risk and safety mean in relation to organisations.

How Can We Know Whether Something Is Safe?

From a historical perspective, humans have always had to worry a little about whether something was safe. It could be whether a bridge was safe to walk on, whether a building was safe or could collapse, whether a journey was safe, whether an activity ('doing something') was safe, etc. While evidence of this concern can be found as far back as in the Code of Hammurabi mentioned above, the problem became more critical after the second industrial revolution and became indispensable after the 'third' revolution

Table 2.1   Technology-based safety questions

Safety concern (question)	Meaning
Design principles	On what basis is the system (machine) designed? Are there explicit and known design principles?
Architecture and components	Do we know what the system components are (what it is made of) and how they are put together?
Models	Do we have explicit models that describe the system and how it functions? Are the models verified or 'proved' to be correct?
Analysis methods	Are there any proved analysis method? Are they commonly accepted or even 'standard'? Are they valid and reliable? Do they have an articulated theoretical basis?
Mode of operation	Can we specify the mode of operation of the system, i.e., is it clear what the system is supposed to do? Does it have a single mode of operation or multiple?
Structural stability	How good is the structural stability of the system – assuming that it is well maintained? Is it robust? Can we ascertain the level of structural stability?
Functional stability	How good is the functional stability of the system? Is its functioning reliable? Can we ascertain the level of functional stability?

– the use of computing technology. Since safety concerns, with good justification, focused on the technology, the answers to these concerns also came from examining the technology. Over the years, a practice was established in which safety issues were addressed by asking the questions shown in Table 2.1. (An actual list of this kind was never produced, but the questions can easily be inferred from the commonly used methods and techniques.)

This set of questions was developed to assess the safety of technological systems and was thus meaningful for technological systems. It was therefore possible to answer the questions, and in that way determine whether or not a system should be considered safe. When the potential contribution of the human factor to safety became an issue, it was natural to take the established approach and try it on the new 'factor'. Due to the urgency of answering the question 'are humans safe?' there was neither time nor opportunity to consider the problem from scratch, so to speak. The same thing happened seven years later, when the concern for the safety of the organisation became the burning issue. But in

Table 2.2   Relevance of common safety questions for technology, human factors and organisations

Safety concern (question)	Answer characteristics for technology	Answer characteristics for human factors	Answer characteristics for organisations
Design principles:	Clear and explicit	Unknown, inferred	High-level, programmatic
Architecture and components:	Known	Partly known, partly unknown	Partly known, partly unknown
Models:	Formal, explicit	Mainly analogies, often oversimplified	Semi-formal,
Analysis methods:	Standardised, validated	Ad hoc, many but unproven	Ad hoc, unproven
Mode of operation:	Well defined (simple)	Vaguely defined, multifarious	Partly defined, multifarious
Structural stability:	High (permanent)	Variable, usually stable but possibility of sudden collapse	Stable (formal organisation), volatile (informal organisation)
Functional stability:	High	Usually reliable	Good, but high hysteresis (lagging)

both cases, the questions were more difficult to answer, as Table 2.2 shows.

So while we can have some confidence in the answers when the safety of technical systems is assessed, we cannot feel the same way when the safety of the human factor or the organisation is assessed. The reason for that is simply that the questions are less meaningful than for technical systems, if not outright meaningless. Yet despite the inability to answer the questions in a meaningful manner, both people (the human factor) and organisations function quite reliably, and certainly reliably enough for modern industrialised societies to function – and, in the case of disasters, to recover. The response to this inability to provide the needed answers has been to increase the efforts, to develop new methods, and to propose intervening hypothetical variables and mechanisms that hopefully will be easier to measure. There is, however, a logical alternative, namely to consider whether the questions might not be the wrong questions to ask. In other words, to ask whether our notion of safety is in fact a reasonable one?

Comments on Chapter 2

The term 'habitat' normally refers to the area or environment in which an organism or ecological community lives or occurs. By extension, it has also been used to describe human-made environments where people can live and work for an extended period of time, for instance an underwater habitat or a space settlement. A socio-technical habitat can be defined as a set of mutually dependent socio-technical systems that is necessary to sustain a range of individual and collective human activities (life and work). While a workplace (such as an office, a hospital, or a factory) can be described as a socio-technical system when it is considered by itself, the sustained functioning always depends on services provided by other socio-technical systems, for instance in relation to transportation of the workforce, distribution of products, communication and control, etc. The combined socio-technical systems constitute a socio-technical habitat.

The description of the three ages of safety is found in Hale, A.R. and Hovden, J. (1998), Management and culture: The third age of safety – A review of approaches to organisational aspects of safety, health and environment, in Feyer, A.M. and Williamson, A. (eds), *Occupational Injury: Risk Prevention and Intervention*, London: Taylor & Francis. The three ages differ in terms of the predominant causes that people would consider and accept, but not really in the understanding of what safety was. Safety was always characterised by the absence of risk and accidents, and the focus was on finding the causes of that. In other words, the three ages favour different causes, but not a different notion of safety. To emphasise this distinction, this book will refer variously to the (three) ages of safety and to the different stages of safety thinking.

It is impossible to overrate the importance of the first book on safety, which was Heinrich, H.W. (1931), *Industrial Accident Prevention*, New York: McGraw-Hill. The book went through four editions over a period of almost thirty years. The background for the book was practical rather than academic, as Heinrich worked as Assistant Superintendent of the Engineering and Inspection Division of Travelers Insurance Company. Since the empirical background was accidents that occurred in the 1920s, care must be taken not to accept everything in the book as fact or as relevant for today's work environments. On the other hand, a closer study of the many cases in the book shows that little has changed in how humans go about accomplishing their work, differences in tasks and technology notwithstanding.

The 'anatomy of an accident' can be found in Green, A.E. (1988), Human factors in industrial risk assessment – some early work, in Goodstein, L.P., Andersen, H.B. and Olsen, S.E. (eds), *Task, Errors and Mental Models*, London: Taylor & Francis. The 'anatomy' is a generic Fault Tree that can be expanded recursively. 'Failure of control', for instance, can be seen as the top event of another 'anatomy' Fault Tree, and so on. In that sense the 'anatomy' is a general fault model.

Julien Offray de La Mettrie (1709–1751) was a French physician and philosopher, who played a role in the French Enlightenment. Among his better-known colleagues were Denis Diderot and Voltaire. Le Mettrie is today mostly known for his work *L'Homme machine* (published in English as *Machine Man*), which proposed the metaphor of the human being as a machine about 200 years before the practical use of digital computers. The suggestion that we could think of the human as a machine has remained with us ever since. In safety it can be found in the characterisation of the human as a 'fallible machine', described in, for instance, Rasmussen, J. (1986), *Information Processing and Human–Machine Interaction*, New York: North-Holland. The 'fallible machine' concept is discussed further in Chapter 3.

The school of thinking commonly referred to as High Reliability Organisations (HRO), started in the late 1980s with the study of high-risk organisations (aircraft carriers, hospitals, nuclear power plants) that somehow succeeded in avoiding the disasters that might have happened. In other words, a study of the absence of failures rather than a study of failures. Karl Weick, already mentioned in Chapter 1, introduced the idea of collective mindfulness as a basis for HRO. See for instance Weick, K.E. and Roberts, K.H. (1993), Collective mind in organizations: Heedful interrelating on flight decks, *Administrative Science Quarterly*, 38, 357–81. There is also a special HRO website at http://high-reliability.org/. Mindfulness is highly relevant to Safety–II, and is referred to several times in the following chapters.

Chapter 3
The Current State

Safety Thinking

The American National Standards Institute defines safety simply as the freedom from unacceptable risk, where unacceptable risk is indirectly defined as a risk for which the probability is too high. Overall this corresponds well to the traditional definition of safety as a condition where nothing goes wrong. We know, of course, that we can never be absolutely certain about what is going to happen, hence that we cannot be absolutely certain that nothing will go wrong. In practice, therefore, being safe means that the likelihood that something can go wrong is acceptably small, so that there is no real need to worry. This is, however, an indirect and somewhat paradoxical definition because it defines safety by its opposite, by the lack of safety and by what happens when it is not present. An interesting consequence of this definition is that safety is also measured indirectly, not by what happens when it is present, or as a quality in itself, but by what happens when it is absent or missing. (This leads to the counterintuitive terminology that a *high* number of adverse outcomes corresponds to a *low* level of safety, and vice versa.)

In relation to human activity it obviously makes good sense to focus on situations where things go wrong, because such situations by definition are unexpected and because they may lead to unintended and unwanted harm or loss of life and property. Although accidents, even spectacular ones (excluding epidemics and war), have happened throughout the history of mankind, the record was sketchy and incomplete until a few centuries ago. One early example is the collapse of the Rialto Bridge in Venice, when it became overloaded with spectators at the wedding of the Marquess of Ferrara in 1444. A bridge collapse is characteristic of the concerns during the first age of safety, which addressed risks related to passive technology and to structures such as buildings, bridges, ships, etc. The concern that technology might fail was increased by the consequences of the second industrial revolution, around 1750, which was marked by the invention of a usable steam engine. While the first steam engines were stationary, inventors soon put them in boats (the first one being in 1763, although the boat sank) or on wheels (a steam road locomotive in 1784 and the first steam railway locomotive in 1804).

The rapid mechanisation of work that followed led to a growing number of hitherto unknown types of accidents, where the common factor was the breakdown, failure, or malfunctioning of active technology. In Chapter 2, this was described in terms of the three ages in safety thinking, where the transition from age to age corresponded to the adoption of a new category of causes (technological, human factors, organisational). What did not change, however, was the basic focus on adverse outcomes, accidents and incidents, and the conviction that safety could be improved by eliminating the causes of accidents and incidents.

Habituation

An unintended but unavoidable consequence of associating safety with things that go wrong is a lack of attention to things that go right. This can be due to practical limitations in terms of time and effort, which means that it is impossible to pay attention to everything – not least in the industrialised societies. In other words, a kind of efficiency–thoroughness trade-off, as is described below. But it can also be the gradual but involuntary decrease in response to a repeated stimulus that is known as habituation. Habituation is a form of adaptive behaviour that can be described scientifically as non-associative learning. Through habituation we learn to disregard things that happen regularly simply because they happen regularly. One formal definition of habituation is that it is a 'response decrement as a result of repeated stimulation'. In academic psychology, habituation has been studied at the level of neuropsychology and has usually been explained on that level as well.

It is, however, entirely possible also to speak about habituation on the level of everyday human behaviour – actions and responses. This was noted as far back as in 1890, when the American philosopher William James (1842–1910), one of the founding fathers of psychology, wrote that 'habit diminishes the conscious attention with which our acts are performed'. This essentially means that we stop paying attention to something once we get used to it – both when it is something that happens (a stimulus) and something that we do. Habituation means not only that we gradually stop noticing that which goes smoothly, but also that we do not think that continuing to notice is necessary. This applies both to actions and their outcomes – and to what we do ourselves as well as to what others do.

From an evolutionary perspective, as well as from the point of view of an efficiency–thoroughness trade-off, habituation makes a lot of sense. While there are good reasons to pay attention to the unexpected and the unusual, it may be a waste of time and effort to pay much attention to that which is common or similar. To quote William James again: 'Habitual actions are certain, and being in no danger of going astray from their end, need no extraneous help'. Reduced attention is precisely what happens when actions regularly produce the intended and expected results and when things 'simply' work. When things go right there is no discernible difference between expected and actual outcomes, hence nothing that attracts attention or leads to an arousal reaction. Nor is there any strong motivation to look into why things went well. Things obviously went well because the system worked as it should and because nothing untoward happened. This means that the technology performed according to specifications, and that people did too. While the first argument – the lack of a noticeable difference between outcomes – is acceptable, the second argument is fatally flawed. The reason for that will become clear in the following.

The Reason Why Things Work

When talking about how work is done at a work place, be it a hospital ward, an aircraft cockpit, a production line, a supermarket, etc., it is common to use the terms 'sharp end' and 'blunt end'.

The sharp end refers to the situation in which work is carried out (treating a patient, monitoring a flight, assembling parts of an engine, scanning purchases at the checkout counter, etc.), which usually also is the place where the consequences of actions show themselves directly and immediately. (Another commonly used term is the 'coal face', referring to coal-mining.) At the sharp we find the people who actually must interact with potentially hazardous processes in order to do their work, regardless of whether they are pilots, physicians, or power plant operators. The blunt end in turn describes the situations and activities that directly or indirectly establish or affect the conditions for what goes on at the sharp end, although they take place away from it. The blunt end is made up of the many layers of the organisation that do not directly participate in what is done at the sharp end, but which still influence the personnel, equipment and general conditions of work at the sharp end. The blunt end stands for the people who affect safety 'through their effect on the constraints and resources acting on the practitioners at the sharp end' via their roles as policy makers, managers, designers, etc. The blunt end is characterised by being removed in time and space from the activities at the sharp end, which unfortunately means that the two are neither calibrated nor synchronised.

Work-As-Imagined and Work-As-Done

When it comes to describing and understanding why things work and why actions succeed, everyone at the sharp end knows that it is only possible to work by continually adjusting what they do to the situation. (In the literature this is commonly described as 'Work-As-Done'.) But the same work looks quite different when seen from the blunt end. Here there is a tendency to emphasise work as it *should* be done (called 'Work-As-Imagined'), given some general assumptions about what working conditions should be like, or are assumed to be. At the sharp end people look at what they do themselves, whereas at the blunt end people look at what others do – or what they assume they should do. (It is ironic that people at the blunt end rarely, if ever, look at what they do themselves, i.e., they do not realise that they are also at a sharp end, although in reality they clearly are.)

When seen from the (traditional) sharp end it is obvious that Work-As-Done is, and must be, different from Work-As-Imagined, simply because it is impossible for those at the blunt end to anticipate all the possible conditions that can exist. Seen from the sharp end it is no surprise that descriptions based on Work-As-Imagined cannot be used in practice and that actual work is different from prescribed work. But this difference is not at all easy to see from the blunt end, partly because it is seen from the outside and from a distance, partly because there is a considerable delay and partly because any data that might exist have been filtered through several organisational layers. Instead, the blunt end blissfully assumes that there neither is nor should be any difference between Work-As-Imagined and Work-As-Done. When a difference between the two is found, it is conveniently used to explain why things went wrong. Because managers rarely look at how they do their own work, and spend all their time 'looking down' rather than 'looking within', this assumption guides their understanding of why adverse events happen and of how safety should be managed.

The justification for this way of thinking is that it works well for technological systems or machines. Indeed, machines – even complicated ones such as nuclear power plants and jet aircraft or even the Large Hadron Collider – work so well because the components work as designed, individually and when assembled into subsystems. We know from a long experience that it is possible to design even extremely complicated systems in every detail and to make certain that they work, by rigorously ensuring that every component functions according to specifications. Machines, furthermore, do not need to adjust their functioning because we take great care to ensure that their working environment is kept stable and that the operating conditions stay within narrow limits. Indeed, it is necessary to do so precisely because machines cannot adjust their functioning, except as a prepared response to clearly specified conditions. And since a human is usually in a system instead of a piece of equipment because it is cheaper or because we might not know how to construct a mechanical component with the same performance, all we need to do is make sure that the human functions like this component.

La Grande Illusion

A technological system, a machine, can be formally described as a 'finite state automaton' or a 'state machine' in terms of a set of inputs, outputs, internal states, and state transitions. But such a description also forces designers (and managers) to think of humans in the same way. In order for the machine to work and produce a pre-defined output, it must get a correct input. This means that the user must respond in a way that corresponds to the pre-defined categories of input. It is also necessary that the output is correctly interpreted by the user, i.e., that it can be correctly mapped onto one of the pre-defined response possibilities. If the user fails to accomplish that, i.e., if the user's response is not included in the expected set of responses, the system will in effect malfunction.

Because of this limitation of technology, designers are forced to consider a finite (and usually rather small) number of interpretations and the reactions that may follow from that. The purpose of the design, combined with suitable training when necessary, is to make the user respond as a finite automaton, and it is not far from that to actually think about the user as an automaton. This obviously goes for what we call human–machine interaction (or user–interface design), whether it is the control room in a chemical plant, an ATM, or a smartphone. But it also goes for social interaction, to the extent that this can be designed. When we send a request for something to be done in an organisation, e.g., for a work order to be issued, we expect that the response will be of a certain kind. The people who receive the request are supposed to respond in a specific and usually rather limited manner, in order for the social interaction to work.

Another justification is found in the undisputed success of Scientific Management Theory. Introduced by the American engineer Frederick Winslow Taylor at the beginning of the twentieth century, Scientific Management had demonstrated how a breakdown of tasks and activities could serve as the basis for improving work efficiency and had by the 1930s established time-and-motion studies as a practical technique. The basic principles of Scientific Management are:

- Analyse the tasks to determine the most efficient performance. The analysis took place by breaking down the tasks into the elementary steps – or even movements – that constituted the building blocks of the activity or work. This was the beginning of what we now call 'task analysis' – either a classic hierarchical task analysis or a cognitive task analysis.
- Select people to achieve the best match between task requirements and capabilities; this in practice meant that people should be neither overqualified nor underqualified.
- Train people to ensure specified performance. This is to ensure that people can do what they have to do, i.e., that they have the necessary minimum of competence, and that they do not do more than they have to do, i.e., that they stay within the boundaries of the task or activity.

A fourth step is to insure compliance by economic incentives, or other kinds of rewards.

The scientific and engineering developments during the twentieth century brought hitherto unheard-of reliability to technology – think only of the computers and cars of the nineteen-fifties compared to the same today. And the belief was that the same kind of reliability could be achieved of human – and a fortiori organisational – functions, and in basically the same way. Scientific Management provided the theoretical and practical foundation for the notion that Work-As-Imagined constituted the necessary and sufficient basis for safe and effective work. (Safety was, however, not an issue considered by Scientific Management and is not even mentioned in Taylor's book of 1911.) The undeniable success of Scientific Management had consequences for how adverse events were studied and for how safety could be improved. Adverse events could be understood by looking at the components and find those that had failed, with the Domino model as a classic example. And safety could be improved by carefully planning work in combination with detailed instructions and training, which served to reduce the variability of human performance. This is recognisable from the widespread belief in the efficacy of procedures and from the equally widespread emphasis on procedure compliance. In short, safety could be achieved by ensuring that Work-As-Done was identical to Work-

As-Imagined. (Scientific Management also extended the first stage of safety thinking (the focus on technology) to humans by treating people as machines or as part of the larger machine. And this was even before human factors as such came into its own.)

The Fallible Machine

Since the traditional human factors perspective is based on comparing humans to machines, it is no surprise that humans have ended up being seen as imprecise, variable and slow. The variability of human performance, meaning that Work-As-Done is different from Work-As-Imagined, has traditionally been seen as a major cause of accidents. This is due to ex post facto reasoning, which says that *if* only person X had done Y' instead of Y, *then* the outcome would have been different (in other words, a counterfactual conditional). In consequence of that, the usual practice of both accident analysis and risk assessment implies a view of safety according to which systems work because the following conditions are met:

- Systems are well designed, meaning that the functions of all the components integrate smoothly to produce the desired outcome. It is in particular assumed that system designers have been able to foresee everything that can go wrong and to take the necessary precautions (such as redundancy, graceful degradation, barriers and defences).
- Systems are built according to the specifications, of the proper materials, and are well maintained. No compromises are made in the choice of components, nor during the building or assembly. Once the system has been built, the performance of components and assemblies is closely monitored and is always up to specifications.
- The procedures that are provided are complete and correct. They are complete in the sense that every conceivable situation or condition is covered by a procedure; and in the sense that each procedure is complete and consistent. (As an aside, the work in a nuclear power plant is estimated to be governed by about 30,000 procedures. In contrast to

that, emergency surgery on the fractured neck of a femur involves a mere 75 clinical guidelines and policies. But even that is far too many.) Procedures are also written in a way that is unambiguous and easy to understand.

- People behave as they are expected to, i.e., as assumed in Work-As-Imagined and, more importantly, as they have been taught or trained to do. This means that there is a perfect match between the instructions and reality, in particular that the training scenarios have a 1:1 correspondence to what actually happens. People are assumed either to be perfectly motivated to Work-as-Imagined, or to be forced to do it in one way or the other.

This way of thinking describes well-tested and well-behaved systems, or in other words, tractable systems (see Chapter 6). It describes the world as seen from the blunt end, many steps removed from the actual workplace – both geographically and in time. In well-tested and well-behaved systems equipment is highly reliable (because it has been designed well and is perfectly maintained); workers and managers alike are ever-vigilant in their testing, observations, procedures, training and operations; the staff are competent, alert, and well trained; the management is enlightened; and good operating procedures are always available in the right form and at the right time. If these assumptions are correct, then any variability in human performance is clearly a liability and the resulting inability to perform in a precise (or machine-like) manner is a disturbance and a potential threat.

According to this way of looking at the world, the logical consequence is to reduce or eliminate performance variability either by standardising work, in the spirit of Scientific Management theory, or by constraining all kinds of performance variability so that efficiency can be maintained and malfunctions or failures avoided. Performance can be constrained in many different ways, some more obvious than others. Examples are rigorous training (drills), barriers and interlocks of various kinds, guidelines and procedures, standardisation of data and interfaces, supervision, rules and regulations.

Right or Wrong?

Categorising outcomes as either successes or failures is, of course, an oversimplification. But it represents the general idea that some outcomes are acceptable whereas others are unacceptable. The criteria for acceptability and unacceptability are rarely crisp and usually vary with the conditions under which work is carried out – either locally (acute or short term) or globally (permanent or longer term). What is good enough in one situation, for instance if there is a significant pressure to complete a task, may not be good enough in another where there is ample time. Yet, given specific circumstances some outcomes will clearly be judged as acceptable and some clearly as unacceptable – possibly with a grey zone between them.

For a given person or a group, a given outcome will be either acceptable or unacceptable but cannot be both (again with the grey zone excepted). At the same time, the outcome of an activity may be judged acceptable by one person and unacceptable by another. The judgement may, furthermore, also change over time due to factors such as hindsight, post-decisional consolidation, etc. So something may be judged as being both right or wrong at the same time, if looked at by people with different values. The difference is, however, not in the outcome itself, in the manifestation, but in the judgement or evaluation of it.

Looking at What Goes Wrong Rather Than Looking at What Goes Right

To illustrate the consequences of looking at what goes wrong rather than looking at what goes right, consider Figure 3.1. This represents the case where the (statistical) probability of a failure is 1 out of 10,000 – usually written as $p = 10^{-4}$. This means that whenever something happens, whenever an activity is started (a patient admitted to an emergency room, a journey begun, a pump started, a valve closed, etc.) then one should expect that it goes wrong in *one* case out of 10,000. (This must, however, be further qualified, since the likelihood must be seen in relation to the typical duration of the event. This duration may be until the patient is discharged from the ER, until the journey has come to an end, until the tank has been emptied – or filled, etc.) The

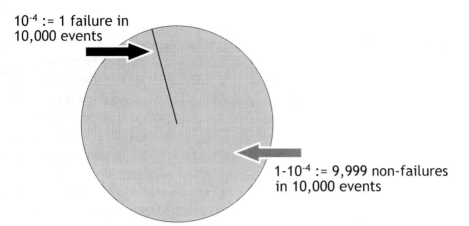

10⁻⁴ := 1 failure in 10,000 events

1-10⁻⁴ := 9,999 non-failures in 10,000 events

Figure 3.1 **The imbalance between things that go right and things that go wrong**

number can, however, also be seen as meaning that for every time we expect that something will go wrong (the thin line in Figure 3.1), then there are 9,999 times where we should expect that things will go right and lead to the outcome we want (the grey area in Figure 3.1). The ratio of 1:10,000 corresponds to a system or organisation where the emphasis is on performance rather than on safety. To give the reader a sense of what 1:10,000 means, the likelihood that an accident would cause damage to a nuclear reactor core (Core Damage Frequency) is 1:20,000 per reactor-year. Compared to that, the likelihood of being in a fatal accident on a commercial flight is 1:7,000,000, which is 350 times smaller. On the other hand, there were 8.99 mishandled bags per 1,000 passengers in 2012 according to the statistics produced by SITA (Société Internationale de Télécommunications Aéronautiques). This corresponds to a likelihood of 1:111. Finally, a ratio of 1:10 represents the likelihood of suffering iatrogenic harm when admitted to a hospital. (The reader may take comfort in the fact that the outcomes – core damage, death mishandled luggage and iatrogenic harm – are practically incomparable.)

The focus on what goes wrong is reinforced in very many ways. Regulators and authorities everywhere require reports on accidents, incidents and even so-called unintended events, and use this to monitor the safety of the system. Most national and international organisations have special offices or units (agencies) that are dedicated to scrutinising adverse events. In this work they can rely on an abundance of models that help to explain how things go wrong, as well as a considerable number of methods that can be used to find and assess the causes. Accident and incident data are collected in numerous databases and illustrated by almost as many graphs; the data are also used by companies to manage the safety of their operations, for instance as inputs to a safety or risk dashboard. Accidents and incidents are described and explained in literally thousands of papers and books, specialised national and international conferences are regularly organised (producing an endless stream of conference proceedings), and national and international research projects are funded to solve the problem once and for all. There is finally a multitude of experts, consultants, and companies that constantly remind us of the need to avoid risks, failures and accidents – and of how their services can help to do just that. The net result is a deluge of information about how things go wrong and about what must be done to prevent these from happening. This focus on failures conforms to our stereotypical understanding of what safety is and how it should be managed. The recipe is the simple principle known as 'find and fix': look for failures and malfunctions, try to find their causes, and try to eliminate causes and/or improve barriers, although this often is adorned by adding more steps and representing it as a development cycle.

The situation is quite different when it comes to that which goes right, i.e., the 9,999 events out of the 10,000. (In the case of aviation, the actual numbers are impressive. In 2012 there were 75 accidents in which 414 people died, but there were also close to 3,000,000,000 flights. So 39,999,999 flights out of 40,000,000 were without accidents.) Despite the crucial importance of these outcomes, they usually receive scant attention. Suggestions to study them further and to understand how they happen generally receive little encouragement. There is first of all no demand from authorities and regulators to look at what works well or to report how often something goes right, and no agencies or departments to do it. Even if someone wanted to do so, it is not easy to know how to do it, and data are therefore difficult to find. There are no readily available methods and only few theories and models.

While we have a rich vocabulary to describe things that go wrong – think of slips, mistakes, violations, non-compliance, plus a rich variety of 'human errors' – there are few terms available to describe that which works. While we have many theories of 'human error' there are few, if any, of everyday human performance. In terms of number of publications, there are only a few books and papers that consider how and why things work, and no specialised journals. And, finally, there are few people or companies that claim expertise in this area or even consider it worthwhile. The situation is a little better when it comes to organisations, since for many years there has been a sustained interest in understanding how organisations function and how they manage to keep free of accidents (cf., the comment on High Reliability Organisations in Chapter 2).

The explanation for this sorry state of affairs is not difficult to find. Looking at things that go right clashes with the traditional focus on failures and therefore receives very little encouragement from the management level. And those who think it is a reasonable endeavour nonetheless have a serious disadvantage when it comes to the practicalities: there are today almost no simple methods or tools and very few good examples to study and learn from.

Safety–I: Avoiding That Things Go Wrong

The argumentation so far can be encapsulated as a particular perspective on safety that I will call Safety–I, for reasons that soon will become obvious. Safety–I defines safety as a condition where the number of adverse outcomes (accidents/incidents/ near misses) is as low as possible. (While 'as low as possible' sounds nice, it really means 'as low as affordable', where what is affordable is determined by considerations of cost, of ethics, of public opinion, etc. 'As low as possible' is therefore not as nice or as simple as it sounds, but that is another story.) It follows from this definition that the purpose of safety management is to achieve and maintain that state, i.e., to reduce the number of adverse events to an acceptable level. Examples of Safety–I are easy to find. A common dictionary definition is that safety is 'the condition of being safe from undergoing or causing hurt, injury, or loss', where 'safe' somewhat redundantly is defined as 'free from harm or risk'. The US Agency for Healthcare Research and Quality defines safety as the 'freedom from accidental injury', while the International Civil Aviation Organization defines safety as 'the state in which harm to persons or of property damage is reduced to, and maintained at or below, an acceptable level through a continuing process of hazard identification and risk management'. The American National Standards Institute similarly define safety as 'freedom from unacceptable risk'. As a consequence of that, safety goals are usually defined in terms of a reduction of the measured outcomes over a given period of time.

The 'philosophy' of Safety–I is illustrated by Figure 3.2. Safety–I promotes a bimodal or binary view of work and activities, according to which the outcome can be acceptable or unacceptable – meaning that the activity can either succeed or fail. This view is reinforced by the standard graphical representation of accidents and risks as a step-by-step development that at each point can branch according to whether the step succeeds or fails. The assumptions in Safety–I is that when the outcomes are acceptable, when the number of adverse events is as low as is reasonably practicable, it is because everything worked as prescribed or as imagined (so-called 'normal' functioning), in particular because people followed procedures. Conversely, when the outcomes are unacceptable, when the results are classified as accidents and

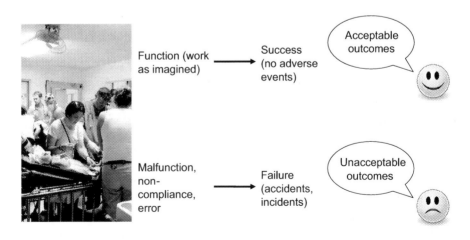

Figure 3.2   Hypothesis of different causes

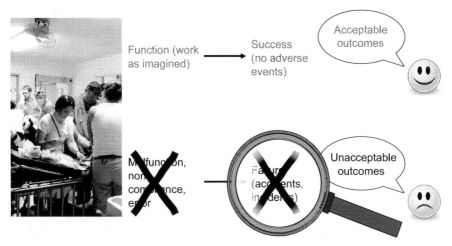

Figure 3.3 **'Find and fix'**

incidents, then it is because something went wrong, because there was a failure or malfunction of a technical or human 'component'.

This view implies that there are two different states of operation of the underlying process or system, one where everything works and one where something fails. The two states are assumed to be clearly different, and the purpose of safety management is to ensure that the system remains in the first state and never ventures into the second. This means there are two ways in which Safety–I can achieve its goals. One is by finding the 'errors' when something has gone wrong and then trying to eliminate them. This is the above-mentioned 'find and fix' approach (Figure 3.3).

The second way for Safety–1 to achieve its goals is by preventing a transition from a 'normal' to an 'abnormal' state (or malfunction), regardless of whether this happens through an abrupt or sudden transition or through a gradual 'drift into failure'. This is done by constraining performance in the 'normal' state, by controlling and reducing its variability (Figure 3.4). It corresponds metaphorically to building a wall around everyday work consisting of various types of barriers: physical barriers that prevent an action from being carried out or the consequences from spreading; functional barriers that hinder an action by means of preconditions (logical, physical, temporal) and interlocks (passwords, synchronisation, locks); symbolic barriers such as signs, signals, alarms and warnings that require an act of

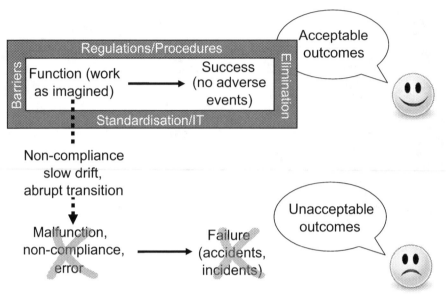

Figure 3.4   Different processes, different outcomes

interpretation to work; and finally incorporeal barriers that are not physically present in the situation but depend on internalised knowledge (e.g., rules, restrictions, laws).

Hypothesis of Different Causes

In addition to trying to prevent a transition from the 'normal' to the 'abnormal' state, a Safety–I perspective also leads to a focus on failures which in turn creates a need to find their causes. When a cause has been found, the next logical step is either to eliminate it or to disable suspected cause–effect links. Since the cause usually can be associated with a specific component, possible solutions are to redesign the process so that the component is eliminated, to replace the component by one that is 'better' – as in replacing humans by automation, to introduce redundancy in the design, to build in further protection or defence in depth, etc. Once this has been achieved, the improved outcome is then measured by counting how many fewer things go wrong after the intervention. Safety–I thus implies what could be called a *hypothesis of different causes*, namely the hypothesis that the causes or 'mechanisms' that lead to adverse outcomes (failures) are different from those

Table 3.1   **Categories of non-compliance pertaining to Work-As-Imagined**

Unintentional non-compliance	Unintentional understanding failure – when people have a different understanding of what the procedure is and what they have to do.
	Unintentional awareness failure – when people are not aware of the existence of a rule or procedure and therefore do not operate with any reference to it.
Intentional non-compliance	Situational non-compliance – when the situation makes it impossible to do the job and be compliant, e.g., because of insufficient time or resources.
	Optimising non-compliance for company benefit – individuals take short cuts believing that this will achieve what they believe the company, and their superiors, really want.
	Optimising non-compliance for personal benefit – short cuts taken to achieve purely personal goals.
	Exceptional non-compliance – deviations from the official procedures that may be difficult to follow under specific, and usually novel, circumstances.

the causes or 'mechanisms' that lead to 'normal' outcomes (successes). If that was not the case, meaning that failures and successes happened in the same way, the elimination of such causes and the neutralisation of such 'mechanisms' would also reduce the likelihood that things could go right, hence be counterproductive.

The 'hypothesis of different causes' is supported by a rich vocabulary for various types of malfunctions and 'errors' – but strangely enough not for successful actions. In the early 1980s there were just two categories, namely 'errors of omission' and 'errors of commission'. But these were soon complemented by various types of 'cognitive errors', deviations and violations. As an illustration of that, consider the following variations of non-compliance in Table 3.1.

Safety management according to a Safety–I perspective can be illustrated as in Figures 3.3 and 3.4. The logic goes through the following steps:

- The system is safe if the number of failures (accidents and incidents) is as low as possible (or affordable).

- Since failures are the results of malfunctions in the underlying process or activity, the system can be safe if the number of malfunctions can be reduced.
- Malfunctions can be reduced by eliminating hazards and risks and by constraining performance to Work-As-Imagined.
- A state of malfunctioning can also be prevented by blocking the transition from a 'normal' to an 'abnormal' state. This can be done by building a 'wall' about the normal functioning, representing the various types of barriers. The 'wall' effectively constraints performance to that which is intended or required, by means of techniques such as barriers, regulations, procedures, standardisation, etc.

The undue optimism of Safety-I in the efficacy of these solution has extended historical roots. Going back ninety years or so, when industrial safety started to come into its own, the optimism was justified since work was relatively uncomplicated, at least compared to the reality today. But as Chapter 6 describes, the work environment has changed so dramatically that the assumptions of yesteryear are invalid today and the optimism in this kind of solution is therefore unfounded.

Safety–I: Reactive Safety Management

The nature of safety management clearly depends on the definition of safety. Since Safety–I is defined as 'a condition where the number of adverse outcomes is as low as possible', the purpose of safety management is to achieve just that. Indeed, safety objectives are usually based on how serious a specific unwanted outcome is, even if the probability of it happening is very low. A good example of that is provided by the cycle shown in Figure 3.5. The figure shows a cyclical repetition of five steps that begins when something has gone wrong so that someone has been harmed. In health care, 'measuring harm' means counting how many patients are harmed or killed and from what type of adverse events. In railways, accidents can be defined as 'employee deaths, disabling injuries and minor injuries, per 200,000 hours worked by the employees of the railway company' or 'train and grade crossing accidents that meet the reporting criteria, per million train miles'. Similar definitions can be found in every domain where safety is a concern.

This approach to safety management is clearly *reactive*, because it starts as a response to something that has gone wrong or has been identified as a risk – as something that could go wrong. The response typically involves trying to find the causes and then to develop an appropriate response (cf., the 'find and fix' solution), such as eliminating the causes, improving options for detection and recovery, reinforcing compliance, introducing new safety measures, etc. This is followed by evaluating the impact – whether the number of accidents or adverse outcomes went down – and finally using the experience to improve safety in the system or process (cf., Figure 3.5).

Examples of reactive safety management are ubiquitous. It is a safe bet that the daily edition of any major newspaper will provide at least one. To illustrate this point, the report into the collision off Lamma Island in Hong Kong on 1 October 2012, was released while this chapter was written. The collision, in which 39 people died, was the worst maritime accident in Hong Kong

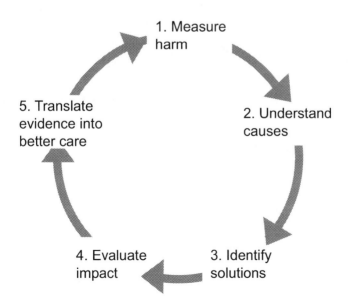

Figure 3.5   **Reactive safety management cycle (WHO)**

for over 40 years. It involved a pleasure boat, the Lamma IV, and a ferry operated by the Hong Kong and Kowloon Ferry (HKKF) company. The investigation into the accident found that 'In the process of designing, constructing and surveying the Lamma IV [...] there was a litany of errors committed at almost every stage by many different people'. In a response to that, the government made it clear that it would take the report seriously and would 'learn the lesson and spare no efforts in making fundamental improvements and reforms to ensure marine safety and restore public confidence'. One might naively ask why this attitude was not present before the accident.

Reactive safety management can work in principle if events do not occur so often that it becomes difficult or impossible to take care of the actual work, i.e., if responding to adverse events does not interfere with the primary activities. But if the frequency of adverse events increases, the need to respond will sooner or later require so much capacity and take so much time that the reactions become inadequate and will partly lag behind the process. Under such conditions the system will not have sufficient time to recover from the response and return to the normal productive state. In practice, it means that control of the situation is lost and with that the ability effectively to manage safety.

Practical examples of this condition are easy to find. Severe weather – tornadoes, typhoons, heavy rain and flooding, or extreme cold or heat – may easily exhaust the capacity of the rescue services to respond. The same goes for forest fires or large oil spills – where the latter can come from ships or from the bottom of the sea. If patients are admitted to hospital for emergency treatment at a rate that is higher than the rate by which they can be treated and discharged, the capacity to treat them will soon be exhausted. This can happen during everyday conditions, or during an epidemic such as the 2006 outbreak of SARS in Hong Kong. On a more mundane level, all regulated industries (power plants, airlines, etc.) are struggling to keep ahead of a maelstrom of incident reports that are mandated by law. Even if only the most serious events are analysed, there may still be insufficient time to understand and respond to what happens.

In order to be effective, reactive safety management also requires that the process being managed is sufficiently familiar and that adverse events are sufficiently regular to allow responses to be prepared ahead of time (anticipation). The worst situation is clearly when something completely unknown happens, since time and resources then must be spent to find out what it is and work out what to do, before a response can actually be given. In order for reactive safety management to be effective, it must be possible to recognise events so quickly that the organisation can initiate a prepared response with minimal delay. The downside of this is that hasty and careless recognition may lead to inappropriate and ineffective responses.

Safety–I is recognisable not only in accident investigation but also in risk analysis and risk management. Since risk analysis looks to possible future events – to something that could happen but has not happened yet, it could be argued that Safety–I in this respect is proactive. Looking ahead is, of course, in principle proactive, but risk analysis is in most cases done only once, at the beginning of a system's life cycle. (In a few safety critical industries it is required to be repeated every n years.) But risk analysis is practically never performed continuously, which would be necessary to qualify as being proactive in practice. Risk management is on the whole reactive rather than proactive because it is a response to risks that have been found through a risk analysis or in the aftermath of an accident.

Safety as a Cost

One unfortunate and counterproductive consequence of the Safety–I perspective is that safety and core business (production) are seen as competing for resources. Investments in safety are seen as necessary but unproductive costs and safety managers may sometimes find it hard to justify or sustain these. The dilemma is captured by the often quoted saying 'If you think safety is expensive, try an accident'. Yet despite this plea, company upper levels and boards often find it hard to understand the importance of investments in safety – particularly if there have been no serious accidents for a period of time or if capital costs have to be recovered. If everything seems to work, then there are no failures. Why, then, invest in preventing them?

This 'accounting attitude' to safety is in some sense justifiable, because investing in increasing compliance and in building additional barriers and defences which do not directly contribute to production certainly is a cost. So the conflict between safety and productivity is real enough in Safety–I. This is one more reason to change the perspective on safety.

Safety–I is reactive and protective because it focuses only on what has gone wrong or could go wrong, and tries to control that, usually by introducing various forms of restriction and compliance. However, since work requires flexibility and variability, protective safety may easily be in conflict with an organisation's attempts to improve productivity. Those responsible for investments that are used to improve safety may sometimes consider whether there will be a negative impact on productivity, but not whether they could actually improve productivity. (The best one can hope for are occasional glorified statements that safety is the highest priority, no matter what the cost. This is in stark contrast to the fact that in times of hardship, safety budgets are the first to be reduced.) Because Safety–I focuses on preventing adverse outcomes, safety competes with productivity. This conflict does not exist in the Safety–II perspective. But before we come to that, it is necessary to consider some of the myths that are used to justify the practices of Safety–I.

Comments on Chapter 3

Habit formation or habituation has been a topic in experimental psychology since the 1920s. The formal definition of habituation used here is from Harris, J.D. (1943), Habituatory response decrement in the intact organism, *Psychological Bulletin*, 40, 385–422. The quotes from William James can be found in James, W. (1890), *The Principles of Psychology*, London: Macmillan and Co. Despite being more than a century old, *The Principles* is an excellent book with many insights that remain valid, often expressed far better than later authors have been able to do.

The sharp end versus blunt end distinction became popular in the early 1990s as a way of illustrating how human performance was determined not only by the current situation but also by what had happened earlier, and in different parts of the organisation.

The focus was mainly on 'human errors', but the blunt end was introduced to make it clear that the 'errors' should not just be blamed on the people who did the work, but also on those who planned, organised and prepared it. The earliest reference is probably in the Foreword to Reason, J.T. (1990), *Human Error*, Cambridge: Cambridge University Press, although the terms here were 'front end' and 'blunt end'. Since sharp end, blunt end thinking requires that some form of linear causality can be established between the two, it is slowly losing its value. As the discussion in this chapter points out, the sharp end and the blunt end clearly have different ideas about how work is done, and this difference is quite important for how safety is managed.

Scientific Management, also known as 'Taylorism' after its main exponent, is a school of thought that aimed to improve labour productivity by a systematic (scientific) analysis of work and work flows. It can be seen as an extension of La Mettrie's thinking about 'man as a machine' to 'work as the performance of a machine', where workers were viewed as replaceable parts of a machine. Taylorism is today often used in a negative sense, as representing a rigid, mechanical approach to work that justifies attempts to reduce Work-As-Done to Work-As-Imagined.

The As-Low-As-Reasonably-Practicable (ALARP) principle is interesting because it directly recognises that safety (in the sense of Safety–I) is a relative rather than an absolute term. This has been expressed very clearly by the UK Offshore Installations Regulations in the following clarification:

> If a measure is practicable and it cannot be shown that the cost of the measure is grossly disproportionate to the benefit gained, then the measure is considered reasonably practicable and should be implemented.

In other words, risks should be as low as an organisation thinks it can afford. The ALARP is another illustration of the counterintuitive terminology, where low risk corresponds to high safety. In Safety–II, the corresponding principle would be AHARP: that safety should be As-High-As-Reasonably-Practicable.

An extensive analysis of the nature and purpose of barriers has been provided in Hollnagel, E. (2004), *Barriers and Accident*

Prevention, Aldershot: Ashgate. This book also provides a more detailed description of the historical development of safety thinking.

Chapter 4
The Myths of Safety–I

Everybody prides themselves on being rational, in the sense that there are reasons for what they do. This is at least so in relation to work, except for the work of artists. The ideal of rational behaviour is that there is a reason for what we do that it is based on an understanding of a situation and a consideration of the pros and cons of a specific choice, at least in cases where decisions are made. (There are clearly large parts of behaviour or performance that are more or less automatic – sometimes called 'skill-based' or 'rule-based' – but even in this case the skills and rules are the encapsulation of the rational choices, which are glorified by the label 'knowledge-based'.) This should in particular apply to performance related to safety, as when we analyse something that has happened or try to anticipate what might happen, especially what may go wrong.

Yet it is also the case that we seem to follow the efficiency–thoroughness trade-off (ETTO) principle in whatever we do. In other words, the actual conditions of work require that we make trade-offs or sacrifices between being thorough and being efficient. This also applies to how we manage safety. Accident investigations always have to be completed before a certain time or within certain resources, the exception perhaps being the extremely rare and extremely serious events. Risk assessments, even major ones, as a rule suffer from restrictions of time and resources. One way this can be seen is that much of the reasoning (or rational thinking?) we are assumed to do, and that we pride ourselves on doing, in fact is based on the mental short cuts that we call assumptions. An assumption is a ready-made conclusion that we trust without going through the required reasoning; it is something that is taken for granted rather than doubted. Assumptions are an important part of human activity and are widely shared among social or professional groups. Assumptions are essential because we never have sufficient time to make sure that what we assume actually *is* true. We take for granted that the assumptions are correct, simply because everybody else uses them. And we trust them for precisely the same reasons.

Assumptions sometimes become so entrenched that they turn into myths. Whereas an assumption is something that is taken for granted without proof, a myth is a belief that has become an integral part of a world view. Assumptions that are sustained for a long time can turn into myths, but by the time this happens the assumptions are probably no longer valid. Assumptions are sometimes questioned – if we have the time and resources to do it – but myths are never questioned, not even if we have the time or resources to do so.

Many people will immediately protest against these remarks and point out that they are well aware that safety management includes both myths and assumptions – in addition to facts and evidence. And it is hardly news that myths are wrong. Indeed, being myths they are almost bound to be. The reason for making these remarks is rather that these myths and assumptions, despite being recognised as such, still are used and still play an important role in safety management. They may not be used by the academic cognoscenti, but they are the bread and butter of how practitioners – for lack of a better term – go about safety and safety management. The myths can be found in, for instance, annual reports, popular beliefs and the general 'knowledge' that everybody has. They thus have a very real effect in practice, even though they ought not to.

Safety–I embodies a number of assumptions and myths and thereby indirectly endorses them. Since these are important determinants of how we perceive adverse outcomes, how we try to understand them and how we respond to them – and, thereby, how we manage safety – it is worthwhile to consider some of the most important ones in more detail.

Causality Credo

The most important of the myths of Safety–I is the unspoken assumption that outcomes can be understood as effects that follow from prior causes. Since that corresponds to a belief – or even a faith – in the laws of causality, it may be called a *causality credo*. This assumption is the very basis for thinking about Safety–I and it can be expressed as follows:

1. Things that go right and things that go wrong both have their causes, but the causes are different. The reason for adverse outcomes (accidents, incidents) is that something has gone wrong (cf., Figure 3.2). Similarly, the reason for successful outcomes is that everything worked as it should, although this is rarely considered.
2. Since adverse outcomes have causes, it must be possible to find these causes provided enough evidence is collected. Once the causes have been found, they can be eliminated, encapsulated, or otherwise neutralised. Doing so will reduce the number of things that go wrong, and hence improve safety.
3. Since all adverse outcomes have a cause (or causes) and since all causes can be found, it follows that all accidents can be prevented. This is the vision of zero accidents or zero harm that many companies covet.

The *causality credo* makes eminent sense in the case when we reason from cause to effect (forwards causality). But it unfortunately tricks us into believing that the opposite, reasoning from effect to cause, can be done with equal justification (backwards causality). While concluding that the antecedent (the cause) is true because the consequent (the effect) is, in most cases may seem to be a plausible inference, it is regrettably logically invalid. Even worse, while it may be plausible for systems that are relatively uncomplicated, it is not plausible for systems that are complicated. This can be illustrated by considering the *causality credo* vis-à-vis the typical accident models.

In accident analysis the law of causality, or rather the law of reverse causality, reigns supreme. Where the law of causality states that every cause has an effect, the reverse law states that every effect has a cause. Although this is reasonable in the sense that it is psychologically unacceptable if events happen by themselves, it is not a defensible position in practice. Furthermore, assuming that a cause must exist does not necessarily mean that it can also be found. The belief that this is possible rests on two assumptions: the law of reverse causality mentioned above; and the assumption that it is logically possible to reason backwards in time from the effect to the cause (the rationality assumption). Quite apart from the fact that humans are notoriously prone to reason in ways that conflict with the rules of logic, the rationality assumption also requires a deterministic world that does not really exist.

When thinking about safety, there must necessarily be a symmetry between the past and the future, which means that future accidents must happen in the same way as past accidents. Or, to put it differently, the reasons why accidents happened in the past must be the same as the reasons why accidents will happen in the future. This does not mean that they are due to the same types of events or conditions, since socio-technical systems constantly change. But it is meant in the sense that the principle of explanation, e.g., the *causality credo*, must be valid for the future as well as for the past. If that were not the case, it would require that some mysterious force in the present – meaning right now! – changed the way things work and the way things happen. Since this is clearly not sensible, the consequence is that there must be a symmetry between that past and the future – and, more specifically, that accident models and risk models should follow the same principles.

It is today common to distinguish among at least three different types of accident models, called the sequential, the epidemiological, and the systemic, respectively. Of these, the sequential and epidemiological comprise the *causality credo*, while the systemic does not.

The prototype for the sequential accident model is the Domino model. The Domino model represents simple, linear causality, as in the case of a set of domino pieces that fall one after the other. According to the logic of this model, the purpose of event analysis is to find the component that failed, by reasoning backwards from the final consequence. This corresponds in all essentials to the principle of the Root Cause Analysis (RCA), which will be described in more detail in the following. Similarly, risk analysis looks for whether something may 'break', meaning that a specific component may fail or malfunction, either by itself or in combination with another failure or malfunction. The combinations are simple logical combinations (AND or OR), and events are assumed to occur in a predefined sequence. Quantitative risk analysis also tries to calculate the probability that a component fails or that a certain combination occurs (Figure 4.1).

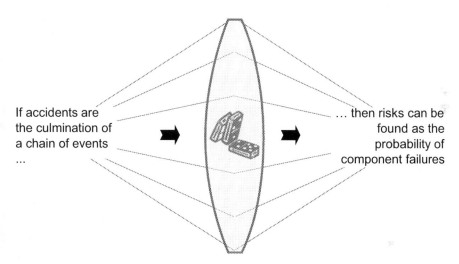

If accidents are the culmination of a chain of events ...

... then risks can be found as the probability of component failures

Figure 4.1   Causes and risks in simple, linear thinking

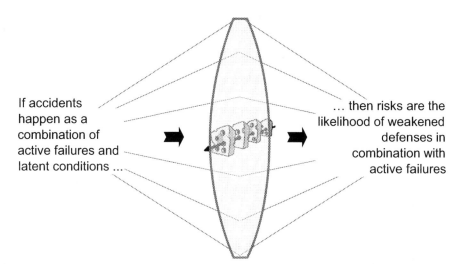

If accidents happen as a combination of active failures and latent conditions ...

... then risks are the likelihood of weakened defenses in combination with active failures

Figure 4.2   Causes and risks in composite, linear thinking

The simple, linear model was superseded in the 1980s by the epidemiological model, the best known example of which is the Swiss cheese model. The Swiss cheese model represents events in terms of composite linear causality, where adverse outcomes are due to combinations of active failures (or unsafe acts) and latent conditions (hazards). Event analysis thus looks for how degraded barriers or defences can combine with active (human) failures. Similarly, risk analysis focuses on finding the conditions under which combinations of single failures and latent conditions may result in an adverse outcome, where the latent conditions are conceived of as degraded barriers or weakened defences (Figure 4.2).

Consequences of the Causality Credo

The *causality credo* also embraces a set of assumptions about how things happen in a system and how systems are composed. The assumptions are as follows:

- The object of study can be decomposed into meaningful elements. These elements can either be components or functions, depending on whether the study looks at a system or at a past or future scenario.
- The performance of each element, be it a component or a function, can be described as if it were bimodal, such as on/off, true/false, works/fails. (It may furthermore be possible to calculate the probability of whether an element will be in one mode or the other, e.g., whether it will work or will fail.)
- In the case of a set of events, such as the parts of an activity, the events will occur in a sequence that is predetermined and fixed. (A good example of that is provided by the THERP tree.) This assumption makes it possible to describe a scenario by means of a branching sequence, i.e., a tree representation. If an alternative sequence is considered, it requires a separate tree representation.
- When combinations of sequences occur, such as in a Fault Tree, they are tractable and non-interacting.
- The influence from the context or the environment is limited and quantifiable. The latter means that the influence can be

accounted for by adjusting the probability of a failure or malfunction.

The Pyramid of Problems

Another myth is that different types of adverse outcomes occur in characteristic ratios, corresponding to different frequencies. The differences in frequency of occurrence are depicted by the so-called 'accident pyramid' or 'Bird triangle', where different types of outcomes correspond to different levels. A typical rendering of the accident pyramid is shown in Figure 4.3. The myth is that the ratios described by the accident pyramid can be used to determine whether the actual distribution of different types of outcomes for a given time period, typically a year, is 'normal' or not. This is sometimes used to determine whether the number of accidents in a company is so high that there is reason to worry or so low that there is no need for concern, for instance as part of an annual report. The interpretation of the pyramid is that the more severe an outcome is, the smaller the number of occurrences will be and, conversely, that the less severe an outcome is, the more occurrences there will be. The numbers shown in Figure 4.3 represent the generally accepted 'reference' ratios.

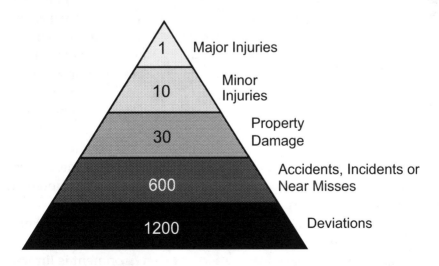

Figure 4.3   The accident pyramid

Origin of the Pyramid

The idea of the accident pyramid is usually attributed to Herbert William Heinrich, one of the true pioneers of industrial safety. The first graphical rendering is found in a paper from 1929, as part of tableau explaining 'the foundation of a major injury'. The graphics showed a small black square representing one major injury; this is placed above a black bar representing 29 minor injuries, which in turn is placed above a thicker and longer black bar representing 300 no-injury accidents (Figure 4.4A). The tableau is accompanied by the following explanation:

> As previously indicated, for each personal injury of any kind (regardless of severity) there are at least ten other accidents; furthermore, because of the relative infrequency of serious injuries, there are 330 accidents which produce only one major injury and 29 minor injuries. In view of these facts, it should be obvious that present-day accident-prevention work is misdirected, because it is based largely upon the analysis of one major injury – the 29 minor injuries are recorded (but seldom analysed) and the 300 other occurrences are, to a great extent, ignored.

The numbers were not pulled out of thin air, but were based on the analysis of 50,000 accidents by the Travelers Insurance Company. They became famous when they were included in the first edition of *Industrial Accident Prevention*, published in 1931, which was the first book to provide a comprehensive description and analysis of accidents. Later studies have tried to confirm the numbers and thereby also the ratios. The most comprehensive was probably a study by Frank Bird in 1969, which analysed 1,753,498 accidents reported by 297 cooperating companies representing 21 different industrial groups. Bird used a set of four outcome categories (fatal accident, serious accident, accident, and incident), and concluded that the ratios were 1:10:30:600.

Heinrich's book went through several editions, and in the fourth edition – published only a few years before Heinrich died in 1962 – the graphical rendering had been changed to the one in Figure 4.4B. Here the square and the three bars are superimposed on what mostly looks like a perspective drawing, showing the 'path' to an injury. The rendering is, however, perceptually ambiguous since it also could be seen as a stylised representation of a pyramid.

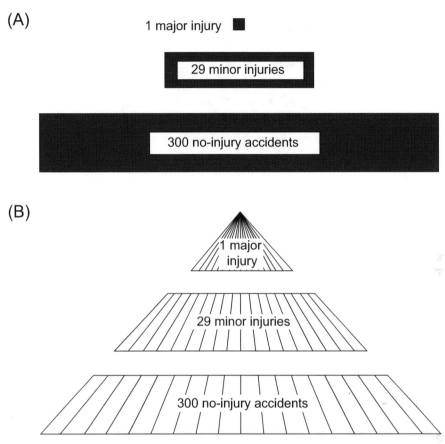

Figure 4.4   Two renderings of the 'accident pyramid'

The accident pyramid is commonly interpreted as suggesting that for every serious major injury there is a larger number of minor injuries, and an even larger number of property damage events and incidents with no visible injury or damage. Since the numbers of events in these categories seem to display a fixed relationship, or at least a relationship that varies little, the accepted assumption is that common causes of accidents exist across the levels and that these follow a logical, linear relationship. According to these assumptions, if more anomalies are identified, then more critical events can be predicted and prevented, which means that more major accidents can be prevented in the end. Life is, however, not that simple.

Problems with the Pyramid

The relationships shown and implied by the accident pyramid are simple, tempting and dubious. The two main problems have to do with the definition of the categories and the implied relationships among the categories.

As far as the definition of the categories is concerned, Heinrich was careful to point out that the labels refer to different types of injury, i.e., of outcomes, but not to different types of accidents. This is clear from the labels he used, shown in Figure 4.4. But in the later versions the levels have become types of events, for instance accident, incident, near miss and unsafe act or (in healthcare) serious accident, serious preventable events, events causing moderate harm, and near misses. Quite apart from a lack of agreement on how the pyramid's levels should be named, none of the many names proposed represent well-defined and unambiguous categories (cf., the discussion about the measurement problem in Chapter 1). While a major injury obviously is more serious than a minor injury, the categories are conveniently assumed to be self-evident. There are therefore no criteria or guidance on how an adverse outcome should be assigned to a specific category, Heinrich's original presentation being no exception.

When Heinrich proposed the three categories in 1929 and counted the number of occurrences, life was relatively simple – at least compared to what it is now. The background was accidents at work in the 1920s, which typically involved a single person (usually male) working with a tool or a machine in industries such as woodworking, metal refining, chemical manufacturing, paper mills, machine shops, hardware manufacturing and iron foundries. From a contemporary perspective these are all quite uncomplicated in the sense that the work processes were relatively independent, with limited horizontal and vertical coupling of work. The injuries were also mainly injuries to the workers at the sharp end, rather than the more dispersed consequences we find today. Indeed, when the accident pyramid is used for today's industries, petrochemical, transportation, aviation, etc., the underlying assumptions may no longer be valid, in particular with regard to the classification of outcomes. A serious accident in the 1990s is vastly different from a serious accident in 1926. In both cases adverse outcomes may be categorised and counted, but the ratio between them has no meaning because the categories are no longer the same.

It always seems to be difficult to maintain a strict separation between that which happens and its consequences. This becomes more of a problem the smaller the consequences are. In the case of an accident, the focus is clearly on the injury or the outcomes, which by definition are non-negligible. But in the case of a near miss the very definition ('a situation where an accident could have happened had there been no timely and effective recovery') means that there will be no consequences, hence nothing to observe or rate. In such cases the label therefore refers to the activity in itself.

The Allure of the Graphical Presentation

Even if we disregard the problems in defining the categories and in assigning outcomes to categories, the proportion or ratio of numbers (quantities) of two categories is only meaningful if there is a systematic relation between them, i.e., if the categories stand in a meaningful relationship to each other. For this to be the case it is necessary that the categories refer to a theory of why the various outcomes happen, specifically how accidents happen. In the absence of that, it makes little sense to use, e.g., an accident-to-incident ratio as a meaningful indicator and 'deduce' that if you reduce the number of incidents you will also reduce the number of accidents.

In his 1929 paper Heinrich was careful to point out that a distinction should be made between accidents and injuries, i.e., between the cause and its effect.

> The expression 'major or minor-accidents' is misleading. In one sense of the word there is no such thing as a major accident. There are major and minor injuries, of course, and it may be said that a major accident is one that produces a major injury. However, the accident and the injury are distinct occurrences; one is the result of the other, and in the continued use of the expression 'major accident', and in the acceptance of its definition as one that results seriously, there is a decided handicap to effective work. In reality, when we so merge the terms 'accident'

and 'injury', we are assuming that no accident is of serious importance unless it produces a serious injury. Yet thousands of accidents having the potential power to produce serious injuries do not so result.

Despite Heinrich's caution, the accident pyramid is today commonly used as if it described different categories of events rather than different categories of outcome or injury. One reason for the different focus could be that today's systems have more layers of protection built in, which means that more events are stopped or caught before they can develop their final consequences. It therefore makes sense to look at the events rather than the outcomes.

Leaving this argument aside, and just taking the categories at face value and as meaningful, the accident pyramid can still be interpreted as representing different relationships among the categories of events/outcomes. I will here present three fundamentally different interpretations; more are, of course, possible. One interpretation is that outcomes represent the cumulated effect of causes and factors, cf. Figure 4.5. The basic idea is that there is a proportionality between causes and effects, so that small causes – just one thing going wrong – correspond to simple consequences, whereas larger causes or a cumulation of causes correspond to serious consequences. (This interpretation is, by the way, congruent to the Domino model, since the final outcome may be prevented if the first cause is.)

Another interpretation is that serious outcomes can be explained by a failure of preventive measures (barriers, defences). In this interpretation, the seriousness of the outcome

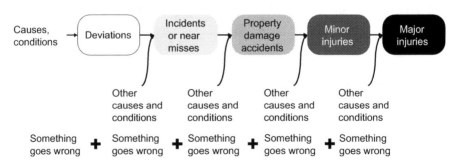

Figure 4.5   Injury categories as cumulated effects

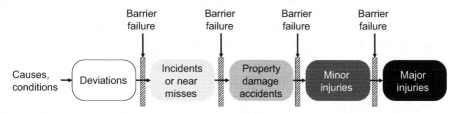

Figure 4.6 **Injury categories as an outcome of compound barrier failures**

depends on the number of barriers that have failed, as shown in Figure 4.6. In this interpretation, different outcomes have a common origin. This could be some kind of harmful influence, the malfunctioning of a piece of equipment or a 'human error' that propagates through the system until it is blocked or reaches the final outcome. Accidents consequently occur if a development goes unchallenged, typically because barriers do not work as expected. (This interpretation is, by the way, congruent to the Swiss cheese model, since the final outcome may be prevented if there are no latent conditions in the system.)

A third interpretation is that the different types of outcomes have different causes, hence that they are independent of each other (cf., Figure 4.7). This means that some causes, alone or in combination, may lead to anything from a 'no outcome' situation ('deviation') to a major injury. Whether one or the other

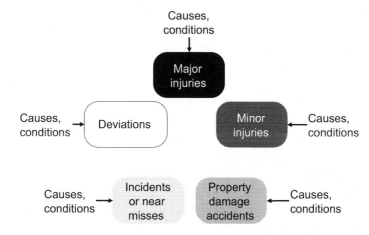

Figure 4.7 **Injury categories as independent outcomes**

happens depends on other things that are not described by the 'model'. Note that if this is the case, the rationale of the accident pyramid disappears. It is still acknowledged that different types of outcomes occur with different frequencies, but that does not imply any kind of underlying 'mechanism'. Each event or outcome must therefore be modelled and analysed separately.

Although the accident pyramid, even in the ambiguous form used by Heinrich (Figure 4.4B), has been interpreted as implying a specific causal relationship or model, this may never have been the intention. The accident pyramid is tempting to use, because it provides a simple representation of an intricate issue. However, the common interpretation of the pyramid rendering suggests something that Heinrich carefully avoided by his rendering. The pyramid implies a vertical dimension, hence a causal relationship between the categories represented by the various levels, but the perspective drawing does not. The interpretation that incidents and near misses potentially lead to major accidents, so that safety management should focus on the former is, therefore, a serious over-interpretation of the graphical rendering and also of what Heinrich actually said. His argument in the 1931 book, based on the actual studies of accidents, was that the difference was in outcomes only.

> The total of 330 accidents all have the same cause. Since it is true that the one major injury may result from the very first accident or from the last or from any of those that intervene, the obvious remedy is to attack all accidents.

The likelihood of misinterpreting the graphical representation may be significantly reduced simply by choosing another rendering. The pie chart in Figure 4.8 represents the same ratios as Figure 4.3, but is less likely to imply a relationship between the three categories.

Finally, to illustrate the dangers in juxtaposing different categories, consider the following facts. In Denmark in 2012, there were 1–2 wolves, 5 bison, and about 100,000 horses. This gives a ratio of 1:5:100,000. But does that mean that the ratio is meaningful? And will reducing the number of bison also reduce the number of wolves?

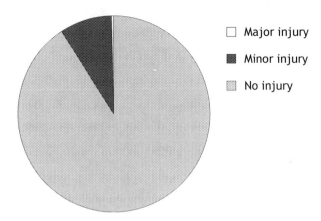

Figure 4.8   The 'accident pyramid' as a pie chart

The 90 Per Cent Solution ('Human Error')

It is a truth universally acknowledged that the overwhelming majority of accidents are due to 'human error'. This is easily illustrated by statements such as 'it is generally recognised that "driver error" is a significant factor in more than 90 per cent of all car accidents', or 'unfortunately, human error is a significant factor in almost all accidents', or 'it has been well documented that human error is a significant factor in maintenance system performance'. (These and thousands of similar statements can easily be found on the web.) The number of books and papers that have been written about 'human error' is prohibitively large and agonisingly repetitive. A few of these try to argue that the universally acknowledged truth is no truth at all, but at best a serious oversimplification and at worst a grave myth. The myth of the 'human error' as the cause of most accidents, however, shows little sign of disappearing.

The use of 'human error' as a possible cause of accidents is old; indeed, one of the early candidates for a theory to explain industrial accidents was a single-factor model of accident proneness put forward in 1919. While accident proneness pointed to humans as potentially unreliable, it did not explicitly describe how an accident happened, i.e., it was a vague assumption rather than a formal model (one reason for that may be that the need for formal accident models in 1910 was rather limited). The first explicit accident model that pointed to 'human error' was in fact the Domino model, which Heinrich described as follows:

> In the middle 1920s a series of theorems were developed which are defined and explained in the following chapter and illustrated by the 'domino sequence'. These theorems show that (1) industrial injuries result only from accidents, (2) accidents are caused directly only by (a) the unsafe acts of persons or (b) exposure to unsafe mechanical conditions, (3) unsafe actions and conditions are caused only by faults of persons, and (4) faults of persons are created by environment or acquired by inheritance.

In the first edition of Heinrich's *Industrial Accident Prevention* published in 1931, the causes of 50,000 accidents leading to injury and 500,000 accidents from which there were no injuries had been tallied with the following results: 'human error': 90 per cent, mechanical hazards: 10 per cent. ('Human error' included the following categories: faulty instruction, inattention, unsafe practice, poor discipline, inability of employee, physical unfitness and mental unfitness.) In other words, the 90 per cent solution.

The concept of 'human error' became part of safety lore when Heinrich noted that as improved equipment and methods were introduced, accidents from purely mechanical or physical causes decreased and (hu)man failure became the predominant cause of injury. This assumption became the second of the five dominoes in the famous Domino model, described as 'fault of person'. This is in good agreement with the philosophical and psychological tradition to treat 'human error' as an individual characteristic or a personality trait. A modern example of that is the zero-risk hypothesis of driving, which proposes that drivers aim to keep their subjectively perceived risk at zero level.

The futility of using 'human error' to explain accidents can be demonstrated by the following argument, cf., Figure 4.9. If we consider a safe system to be one where the probability of failure is low, e.g., 10^{-5}, then there will be at least 99,999 cases of acceptable performance for every case of unacceptable performance. In other words, accidents will be quite rare. But if the so-called 'human error' is the cause of the event that goes wrong, then what is the cause of all the other events that go right? The only reasonable answer is humans: humans try to make sure that their actions

When something goes wrong, e.g., 1 event out of 100,000 (10E–5), humans are assumed to be responsible in 80–90 percent of the cases.

When something goes right, e.g., 99,999 events out of 100,000, are humans also responsible in 80–90 per cent of the cases?

Who or what are responsible for the remaining 10–20 per cent?

In these cases, an investigation of failures is important.

Who or what are responsible for the remaining 10–20 per cent?

In these cases, an Investigation of successes is rarely undertaken.

Figure 4.9   The dilemma of 'human error'

produce the intended effect. However, they behave in the same manner regardless of whether the outcome of their actions turns out to be positive or negative, simply because they cannot know that at the time of acting. It follows that 'human error' should not be used to explain adverse outcomes since it requires an ad hoc 'mental mechanism'. Instead, a more productive view is to try to understand how performance varies and determine why the behaviours that usually make things go right occasionally make things go wrong.

The practical basis for the utility of 'human error' is that people regularly do things that in one way or another are wrong and which may lead to unwanted consequences. The 'wrongness' may be subjective and therefore only noticed by the acting person, or intersubjective and noticed by others as well. Each of us will several times a day experience that we have done something we did not intend to do or not done something we intended to do. Most of these cases concern trivial matters, such as making a typing error or forgetting to buy something at the supermarket, but in some cases the slip-up may seriously affect what we are doing and perhaps even have consequences for others. The 'slips and lapses' of everyday actions are easily explained in a number of ways, such as lack of experience, confusing work layout, conflicting demands, insufficient information, etc. In most cases people fortunately realise what has happened in time to recover from the failure and therefore avoid any significant consequences.

As the utility of 'human error' lies in the description of the ways in which something can be done wrongly, the issue is how best to describe the manifestations of 'human error' or 'error-as-action'. There is an indisputable need for a practical vocabulary in order to be able to describe such failures in a consistent manner. The term 'human error' is, however, of little use both because it is a catch-all category and because it confuses actions and causes. What is needed is actually not a theory of 'human error' as such, but rather a consistent classification of manifestations. Fortunately, such a classification is not hard to come by. The best-known example is the traditional Hazard and Operability (HAZOP) technique, which in the early 1960s came into use in the engineering disciplines as a way of analysing all the possible outcomes of equipment failure analysis. The basis of a HAZOP is a set of guide words that help the analyst identify all the possible ways in which a function can fail. The guide words are juxtaposed with all significant functions in order to prompt the analyst to consider whether such a combination could occur. The guide words that are used are 'no', 'less', 'more', 'as well as', 'other than', 'reverse' and 'part of'. If, for instance, the function is the opening of a valve, the HAZOP procedure demands that the analyst consider whether 'not opening', 'opening less', 'opening more', etc. may possibly happen.

An unwanted side effect of using 'human error' to explain accidents is that the level of 'human error' becomes the maximum depth of analysis or the universal root cause. Accident investigations often seem to assume that the processes as such were infallible, or would have been so had the operators not done something wrong. It therefore becomes 'natural' to stop the analyses once a 'human error' had been found. This was elegantly expressed by Charles Perrow in 1984, when he wrote:

> Formal accident investigations usually start with an assumption that the operator must have failed, and if this attribution can be made, that is the end of serious inquiry. Finding that faulty designs were responsible would entail enormous shutdown and retrofitting costs; finding that management was responsible would

threaten those in charge, but finding that operators were responsible preserves the system, with some soporific injunctions about better training.

The Last Train from Stockholm

During the early hours of Tuesday 15 January 2013, a rather unusual train accident happened on Saltsjöbanan, an 18.5-kilometre electrified suburban rail system between Stockholm and Saltsjöbaden in the municipality of Nacka, Sweden. The last train of the day train arrived at the depot in Neglinge at 01:45 in the morning. On board was a train operator (a shunter), responsible for driving the train to its final position, and a female cleaner. The next thing that is known is that the train left the depot at 02:23, still with the female cleaner on board, and drove about 2.2 km to Saltsjöbaden. This is the last station on the line and the tracks stop there. It was afterwards determined that the train had been going at high speed, estimated to about 80 km/h, for the last 1.5 km. When it approached the last stop at around 02:30 it did not slow down, but instead drove straight through the so-called buffer stop and ran into an apartment block about 50 metres away. One of the carriages was suspended in mid-air.

Miraculously, no one in the house was harmed, although the ground floor was badly damaged. The tenants of the building were all evacuated. The 22-year old woman cleaner was found right behind the driver's cabin with injuries to both legs, fractures of the pelvis, nine broken ribs, one punctured lung and a half-torn ear. It took more than two hours to free her from the train wreckage, after which she was flown by helicopter to the Karolinska University Hospital in Stockholm. Here she was treated and was kept sedated for three days.

This was clearly an event that should not have happened (as the CEO of the train company immediately made clear). It was, however, an undeniable fact that the train had started to move when it should not. Since trains are not supposed to do that and since the female cleaner was found in the train, the 'obvious' conclusion was that for some reason she had started the train and then possibly lost control of it. In other words, the cause would seem to be a 'human error', possibly a violation of some kind. (In the immediate commotion after the accident, while the woman was still lying unconscious at the hospital, she was even accused of having tried to steal the train, although the mind boggles at what she would have done with it. The accusation was soon withdrawn with many apologies, but it was still indicative of a general mindset.)

In April 2013, the female cleaner gave her account of what had happened in the weekly magazine published by her union. Her job was to clean the trains at the end of the day. She usually did it together with a colleague, who, however, had called in ill on the night. She therefore had to work harder, and had no time to talk to the shunters. During the cleaning, she noticed that the train was moving. This was not unusual since it happens that the trains moved in the pits. But suddenly she had the feeling that it was going too fast. She seemed to remember that she tried to turn the key (in the driver's cabin) to stop the train, but that it was impossible. When she realised that the situation was becoming very serious – the uncontrollable train running through the night at high speed – she found a place where she sat down and braced. Seconds later the train crashed into a residential building in Saltsjöbaden, and three days afterwards she woke up in the hospital.

This account perfectly explains why she was found in the front of the train after the crash. Indeed, it must be considered a natural reaction to try to stop the train. But it does not answer the question of why the train started to move. The investigation has not been completed at the time of writing this chapter, but a number of possibilities have begun to appear in the press. It is known is that the switch was in the wrong position and that the train was ready to run. There is also evidence to suggest that what is called 'the dead-man's switch' was disconnected to prevent the brakes from freezing during the night. Exactly what caused the train to start rolling is still not known, but five different investigations are ongoing. And the public transport commissioner at the Stockholm county council has rather predictably ordered a review of 'security routines'. *Plus ça change, plus c'est la même chose.*

Problems With the 90 Per Cent Solution

There are several problems with having 'human error' as the predominant solution to things that go wrong. Methodologically, it is rather suspicious if one type of cause can be used to explain 90 per cent of all accidents. If it were true, which it is not, then it would at the very least suggest that there was something fundamentally wrong with the way socio-technical systems were designed and operated, not least because the 90 per cent solution has been applied for many decades. Another problem is the lack of agreement about what a 'human error' really is, which is due to different premises or different points of departure. To an engineer, humans are system components whose successes and failures can be described in the same way as for equipment. To psychologists, human behaviour is essentially purposive and can only be fully understood with reference to subjective goals and intentions. Finally, to sociologists 'human errors' are due to features such as management style and organisational structure, often seen as the mediating variables influencing error rates. A third problem is that 'human error' is not a meaningful term at all because it has more than one meaning. It can be used to denote either the *cause* of something, the *event* itself (the action), or the *outcome* of the action.

In recent years, the greatest concern about the 90 per cent solution has been that blaming people is both ineffective and counterproductive (cf., the quote from Charles Perrow in the preceding section). It is ineffective because the standard recommendations, such as 'improve procedures' or 'do not deviate from the instructions', have little discernible effect. And it is counterproductive because blaming people reduces their willingness to cooperate, both in the investigations and in general. Several leading safety experts have pointed out that the greatest single impediment to error prevention is that people are punished for making mistakes.

The dilemma can be explained as follows. Safety–I requires that adverse outcomes are investigated so that they can be effectively prevented. There is a need to learn from accidents and incidents in order to be able to take appropriate action. Since, furthermore, it is assumed that humans play a critical role in most of these cases, humans become an essential source of information. In the current paradigms, there is a need to know what people did and what they thought, what they paid attention to and how they made their decisions. Since very little of this can be recorded or even observed, investigators rely on the willingness of people to provide this information. But this cannot be done if people fear that they may be incriminated or held responsible for everything they did.

As a solution to this self-created problem, it has been proposed that the organisation must embrace a just culture, with aviation and health care being among the leaders. A just culture has been described as 'an atmosphere of trust in which people are encouraged, even rewarded, for providing essential safety-related information – but in which they are also clear about where the line must be drawn between acceptable and unacceptable behaviour'. This requires a delicate balance, since on the one hand it is important not to blame and punish people for what they have done if it is commensurate with their training and experience; but on the other hand it is also important not to tolerate gross negligence, wilful violations and outright destructive acts. The 90 per cent solution thus creates a number of problems, which may disappear if the 90 per cent solution is abandoned.

Root Causes – The Definitive Answer

The natural consequence of the *causality credo*, combined with the Domino model, is the assumption that there is a basic or first cause, which can be found if the systematic search is continued until it can go no further. This is often called the 'root cause', although definitions differ. In the Domino model, the root cause was the 'ancestry and social environment', which led to 'undesirable traits of character'. Since this was the fifth domino, it was not possible to continue the analysis any further. Other approaches, particular if they subscribe to some form of abstraction hierarchy, suffer from the same limitation. The type of analysis (which actually is a family of methods) that tries to find the root cause is unsurprisingly called Root Cause Analysis (RCS).

The goal of a root cause analysis is, sensibly enough, to find out what happened, why it happened, and what can be done to prevent it from happening again. In order to achieve this, the method explicitly makes a number of assumptions: first, that each failure has one (or more) root causes, barring random fluctuations; second, that if the root causes are eliminated, then failure is rendered impossible; third, that the system can be analysed by decomposing it into basic elements; and fourth, that the dynamic behaviour of the system can be explained on the level of the dynamics of the decomposed system elements.

The Shattered Dream

The search for a root cause often meets with success. A sterling example is the blackout in New York on 14 August 2003, which started when a system-monitoring tool in Ohio became ineffective because of inaccurate data input. In this case the analysis was obviously helped by the fact that the electrical grid has a clearly defined structure, which makes it feasible to trace effects back to their causes. But in other cases it may be more difficult, as the following example illustrates.

In May 2006, the Boeing company announced its new 'game change' aircraft, the 787 Dreamliner. Two of the innovative characteristics of the Dreamliner are the use of composite materials as the primary material in the construction of its airframe; and the replacement of conventional hydraulic power sources with electrically powered compressors and pumps, thus completely eliminating pneumatics and hydraulics from several subsystems.

The Dreamliner was initially expected to go into service in 2008. The production of the plane was, however, more complicated than originally thought, and there were a number of serious delays in delivery. The general manager of the 787 programme at one stage remarked that every new aircraft development programme was like a 'voyage of discovery', with unexpected problems and hiccups. In October 2007 Boeing announced that it would delay initial deliveries to late November or early December 2008, because of difficulties with the global chain of suppliers, as well as unanticipated difficulties in its flight-control software. In December 2008 it was announced that the first deliveries would be delayed until the first quarter of 2010, later revised to 2011. The first 787 Dreamliner was officially delivered to All Nippon Airways on 25 September 2011, nearly two years behind the original schedule.

By the end of 2012, 49 aircraft had been delivered to various airlines and taken into operation. During December 2012 and

January 2013, a number of aircraft experienced problems with their batteries, which were damaged or caught fire. On 7 January, a battery overheated and started a fire on a Japan Airlines 787 at Boston's Logan International Airport. On 16 January, an All Nippon Airways 787 had to make an emergency landing in Japan after a battery started to give off smoke. On the same day, JAL and ANA announced that they were voluntarily grounding their fleets of 787s after multiple incidents, including emergency landings. Following several similar events in the US, the Federal Aviation Administration (the national aviation authority of the US) issued an emergency airworthiness directive on 16 January, 2013 ordering all US-based airlines to ground their Boeing 787s. Regulators around the world (Europe, Africa, India and Japan) followed, and soon the 787 was grounded throughout the world.

Following the tradition of Safety–I, a concentrated effort by Boeing, as well as by the airlines, started to look for the possible causes of the 'burning batteries'. This would seem to be a relatively straightforward thing to do, since the problem occurred in what must be considered a purely technical system. The two lithium-ion batteries in question are actually not used when the 787 is in flight but power the aircraft's brakes and lights when it is on the ground and the engines are not running. This use is completely automatic, controlled by the technology itself without depending on human intervention or interaction. Humans are, of course, involved in the design, manufacturing, installation, and maintenance of the batteries, and this was included in the set of possible causes. But the problems were fundamentally technical, which made it reasonable to assume that they could be analysed to the bottom and that a definitive cause – or a set of causes – be found. That this failed was not for a lack of trying. According to Boeing, more than 500 engineers worked with outside experts to understand what might have caused the batteries to overheat, spending more than 200,000 hours of analysis, engineering work and tests. Despite that, the Dreamliner's chief engineer admitted in an interview in March 2013 that they might never know what caused the battery malfunctions. The general manager of the 787 programme added that since it was not uncommon that a single root cause could not be found (sic!), the industry best practice was to address all of the potential causes, and that this is what Boeing was going to do. Root cause, RIP.

Since Safety–I requires that a cause can be found and that something can be done about it, the inability to find a root cause, or set of root causes, was a serious setback. (In addition, it was a serious problem financially for both the Boeing company and the airlines.) The freedom of 'unacceptable risk' can best be achieved by eliminating the risks and hazards. If that is not possible, the second best solution is to introduce additional barriers so that any consequences of an accident, should it happen again, will be limited. This is what has happened in the case of the Dreamliner. Improved batteries have been introduced, that do not have to work so hard, and therefore operate at a lower temperature. In the new design, the batteries are also enclosed in stainless steel boxes which have a ventilation pipe that directly goes to the outside of the plane. According to Boeing, any future 'rare cases' of battery failure will be '100% contained', and any smoke will immediately leave the plane. (As an aside, Boeing had initially estimated that the probability of a battery failure incident was one per 10 million flight hours. In reality there were two separate cases of such incidents in the first 52,000 flight hours. Were problems to continue at the same rate, that would correspond to 384 failure incidents per 10 million flight hours.)

At the time of writing this chapter, the 787 had been given permission to fly again in the US and Japan. On 12 July 2013, a fire on a Dreamliner parked in Heathrow airport created a scare. According to the initial investigation by the UK Air Accidents Investigation Branch, the fire was in the upper rear part of the 787 Dreamliner, where the Emergency Locator Transmitter (ELT) is fitted, hence unrelated to the problems with the lithium batteries.

Problems with Root Cause Analysis

Root cause analysis is attractive because it promises to provide a simple but definitive answer to a problem. It thus satisfies what Friedrich Nietzsche called a 'fundamental instinct to get rid of (these) painful circumstances'. It is therefore hardly surprising that this set of methods is widely applied in many industries, for instance in health care. Indeed, the first response to a major accident or a major problem is often a promise to find the root cause ('any explanation is better than none at all' – as Nietzsche also said). This habit seems to be especially prevalent at the blunt end, including CEOs, military leaders and heads of state, who typically are not accountable for what happens later on when public concern has subsided.

A major problem with root cause analysis as well as most other methods used by Safety–I is that providing a definite answer rules out any alternative explanations or even any motivation to search for 'second stories'. This is serious, not least because of the 90 per cent solution, the predilection for 'human error' as a cause of accidents. Taken together, the myth about the 90 per cent solution and the myth about the root cause mean that many investigations conclude by pointing at the human as the 'root cause'. The problem is exacerbated by the common assumption (which more properly should be called a delusion) that accident investigations represent rigorous reasoning guided by logic and unaffected by practical issues, personal biases, or political priorities. We like to assume that because it puts accident investigations on an equal footing with scientific investigations in general. Looking for a 'root cause' is a legitimate concern in practically all fields of science, for instance cosmology (the Big Bang), physics (the Higgs boson), medicine (the cause of specific diseases, for instance, cancer), psychology (the nature of consciousness), and so on. So why should it not be the same for safety management? By making such comparisons we forget (or wilfully deny) that looking for explanations of events in socio-technical systems is more like interpreting a text (hermeneutics) than analysing a material to determine its chemical composition. It is a psychological rather than a logical process, and therefore does not have a unique solution. While a search for the root cause may be efficient – in the short run, at least – it can never be thorough.

Other Myths

The four myths presented here (the *causality credo*, the accident pyramid, the 90 per cent solution, and the root cause) are of course not the only ones. Readers who are interested in other myths can quite easily find information in scientific papers and online encyclopedias. Other common myths are: the belief in compliance – that systems will be safe if people stick to the procedures; the belief in the defence and protection – that increasing the number and depth of barriers will lead to higher safety; the belief that all accidents are preventable – all accidents have causes, and causes can be found and eliminated; the belief in the power of accident investigations – that accident investigations are the logical and rational identification of causes based on facts; the belief in the congruence between causes and outcomes – that large and serious outcomes have large and serious causes and vice versa; the belief in safety culture – that higher levels of safety culture will improve safety performance; and the belief in 'safety first' – that safety always has the highest priority and that safety is non-negotiable.

While it would be nice if the myths could be debunked, that is probably not going to happen. As is noted in the beginning of this chapter, we already know that the myths are wrong and that usually they are more a hindrance than a help. The way forward is rather to expose them, to look at them with fresh eyes and thereby realise, like the child in Hans Christian Andersen's tale about the Emperor's New Clothes, that there is nothing there. Since the myths are an integral part of Safety–I, it may be difficult to do so from the inside, so to speak. Adopting a Safety–II perspective, not as a replacement of Safety–I but as a complement to it, may make this a bit easier.

Comments on Chapter 4

In the 1970s it became popular to describe humans as information processing systems or machines, and to explain various forms of observable performance in terms of different types of 'mental' information processing. (As in the case of Taylorism, this can be seen as a reincarnation of La Mettrie's idea from 1748.) One of the more successful proposals was to distinguish between three types of processing, called skill-based, rule-based and knowledge-based, respectively. This idea was presented in Rasmussen, J. (1979), *On the Structure of Knowledge – A Morphology of Mental Models in a Man–Machine System Context* (Risø-M-2192), Denmark: Risø National Labs, and soon became very popular, among other things as the foundation for models of 'human error'. It was widely used in the 1980s and 1990s, but is no longer considered adequate. The same source also includes one of the early descriptions of the abstraction hierarchy that is referred to later in the chapter. This specific abstraction hierarchy is a system description based on five levels, from physical form to functional purpose, where each level is proposed as an 'abstraction' of the one below. Abstraction hierarchies are often used as the theoretical justification of accident analysis methods.

Causality is an issue that always has interested philosophers – from Aristotle and onwards. The Scottish philosopher David Hume (1711–1776) is known for his analysis of causation and for pointing out that while cause and effect are observable (physical), causation is not observable (hence metaphysical). On this, a later philosopher, Charles Sanders Peirce (1839–1914), offered the following advice:

> metaphysics is a subject much more curious than useful, the knowledge of which, like that of a sunken reef, serves chiefly to enable us to keep clear of it.

A description of the three types of accident models can be found in Hollnagel, E. (2004), *Barriers and Accident Prevention*, Aldershot: Ashgate. As already mentioned in the Comments on Chapter 3, the book also provides a detailed analysis and characterisation of various barrier systems. The best known among the many accident models are the Domino model put forward by Heinrich in 1931, and the Swiss cheese model described by Reason, J. (1990), *Human Error*, Cambridge: Cambridge University Press.

A THERP tree is a technique used in human reliability assessment to calculate the probability of a human error during the execution of a task. (THERP stands for Technique for Human Error Rate Prediction.) A THERP tree is basically an event tree, where the root is the initiating event and the leaves are the possible outcomes. THERP is described in a publication from 1983 (Swain, A.D. and Guttmann, H.E., *Handbook of Human Reliability Analysis with Emphasis on Nuclear Power Plant Applications*, NUREG/CR-1278, USNRC), and is still widely used despite its unrealistic assumptions about human performance. One important contribution of THERP was the notion of performance-shaping factors, which still is a hotly debated issue.

The accident pyramid appeared in 1931 in Heinrich's book, although in the rendering shown in Figure 4.4A, rather than as a pyramid. In the book it was called 'a chart showing frequency of accidents by degree of severity'. Data from the study by Frank Bird can be found in Bird, F.E. and Germain, G.L. (1992), *Practical Loss Control Leadership*. Loganville, GA: International Loss Control Institute. The contemporary criticism of the pyramid model was probably started by Hale, A., Guldenmund, F. and Bellamy, L. (2000), *Focussed Auditing of Major Hazard Management Systems*, which was presented at the ESReDA (European Safety, Reliability & Data Association) conference held in Karlstad, Sweden on 15–16 June 2000.

The quote by Charles Perrow is from his seminal book *Normal Accidents*, published in 1984 (New York: Basic Books). The many scientific and practical issues with 'human error' have been discussed in Hollnagel, E. (1998), *Cognitive Reliability and Error Analysis Method*, Oxford: Elsevier Science Ltd. The definition of just culture can be found at the *skybrary* (http://www.skybrary. aero), an excellent source of aviation safety knowledge that is also of considerable general interest.

An analysis of some of the other myths can be found in Besnard, D. and Hollnagel, E. (2012), I want to believe: Some myths about the management of industrial safety, *Cognition, Technology & Work*, 12. Another example of relevance to this chapter is Manuele, F.E. (2011), Reviewing Heinrich: Dislodging two myths from the practice of safety, *Professional Safety*, October, 52–61.

Chapter 5
The Deconstruction of Safety–I

Safety Is Safety Is Safety Is Safety

Having so far outlined the history of safety as well as characterised the principles and practices of safety as they can be seen from daily practices, the time has come for a more formal and systematic analysis of Safety–I. This is needed in order fully to account for the strengths and weaknesses of the Safety–I perspective, and through that also to provide the foundation for a complementary view – Safety–II. This will be done by performing a deconstruction of Safety–I.

Deconstruction is originally the name of a philosophical movement which promotes a theory of literary criticism or semiotic analysis. It asserts that words can only refer to other words, and that there is no meaning to be found in an actual text. The meaning is instead constructed by the reader in his or her search for meaning. In the current context, deconstruction is not used in the original sense, but more as illustrating a principled approach to finding what the meaning of safety is as a phenomenon and what the basis for that meaning is. (The reader may recognise that deconstruction also can be used to characterise the ways in which accident investigations are carried out as a search for meaning, in this case for the cause or the explanation.) The deconstruction starts with the phenomenon (in this case the typical safety practices – or, rather, Safety–I practices) and develops an explanation for why certain manifestations are considered meaningful (for a safety manager, for instance). Rather than assuming a priori that the term, or concept, of safety is meaningful, a deconstructionist analysis tries to determine where the meaning comes from and to clarify what the necessary assumptions are.

The deconstruction of Safety–I – which means of safety as we usually think of it – will be done in three steps, corresponding to three different aspects of safety, cf., Figure 5.1.

- The first step addresses the *phenomenology* of safety. The phenomenology refers to the observable characteristics or the manifestations of safety, in other words what it is that makes us say that something is safe – or unsafe. (The word 'phenomenology' is of Greek origin and literally means 'the study of that which appears'.) The first step can also be said to consider the *safety phenotype*, meaning the observable characteristics or traits that we associate with safety.
- The second step addresses the *aetiology* of safety. Aetiology is the study of causation, of why things occur, or even the study of the reasons or causes behind what happens. In medicine, aetiology is the study of the causes of diseases. In relation to safety it is the study of the (assumed) reasons for or causes of the observable phenomena. The aetiology describes the 'mechanisms' that produce the observable phenomena, the adverse outcomes that constitute the focus for Safety–I. The second step can be said to consider the *genotype* of safety, namely that which explains the manifestations or phenotypes. Root cause analysis, described in Chapter 4, is a good example of that.
- The third step addresses the *ontology* of safety. The ontology studies that 'which is', which in this context would be the true nature and the essential characteristics of safety. It is the study of what safety *is*, rather than how it manifests itself (the phenomenology) or how these manifestations occur (the aetiology). It is the question of how we can – or should – understand why safety manifests itself in a certain way, hence in a sense a question of the very 'root' of our understanding of safety. Ontology, in short, is about what *really* goes on.

Deconstruction should not be confused with decomposition, which simply means that something – a system – is broken into

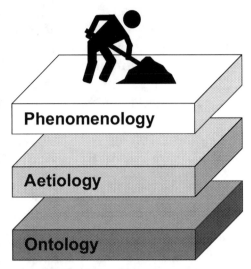

The observable characteristics or manifestations.
The safety phenotype.

The origin or causes of the observable phenomena.
The safety genotype.

The nature and essential characteristics of safety.
What 'really' goes on.

Figure 5.1   The deconstruction of safety

its constituents or parts. Decomposition is the most common approach in science, and is used to explain something by referring to characteristics of the constituents or parts, i.e., by reducing something to something simpler. The study of elementary particles is an excellent example of that. Decomposition is very similar to reductionism, which holds that a system is nothing more than the sum of its parts, and that an account of the system therefore can be reduced to accounts of individual constituents and how they interact. The principle of reductionism can be applied to objects, phenomena, explanations, theories and meanings. We tend, for instance, to explain diseases by looking for genes, or explain consciousness (or life) by referring to neural processes or processes in the brain. Reductionism also plays a role in the discussions of resultant versus emergent phenomena, cf., Chapter 7.

The Phenomenology of Safety–I

Safety, in the guise of Safety–I, is defined as a condition where the number of adverse outcomes (accidents/incidents/near misses) is as low as possible. It follows from this definition that the phenomenology of Safety–I is the occurrence of adverse outcomes. Since adverse outcomes are assumed to be the results of adverse events (failures and malfunctions), the phenomenology is extended from the outcomes alone to include also the cases or situations where things go wrong, i.e., the events themselves. The UK National Health Service, for instance, defines an incident as 'any unintended or unexpected incident which could have or did lead to harm for one or more patients receiving NHS care'. In other words, there was a potential for harm but not necessarily any actual adverse outcome.

The manifestations of Safety–I are accidents, incidents, near misses, etc., as illustrated by the different 'levels' of the safety pyramid or by lists proposed by specific safety programmes, such as the European Technology Platform on Industrial Safety (ETPIS) described above. We thus say that a system is unsafe if such events happen, particularly if accidents happen. Conversely we say that it is safe, if no such events happen – remembering Karl Weick's definition of safety as 'a dynamic non-event', discussed earlier.

The paradoxical consequence of this definition is that the level of safety is inversely related to the number of adverse outcomes. If many things go wrong, the level of safety is said to be low; but if few things go wrong, the level of safety is said to be high. In other words, the more manifestations there are, the less safety there is and vice versa. A perfect level of safety means that there are no adverse outcomes, hence nothing to measure. This unfortunately makes it impossible to demonstrate that efforts to improve safety have any results. The same problem exists for stable levels of safety, meaning a condition where the number of adverse outcomes is constant. Here it is also impossible to demonstrate any effects of interventions or efforts. In both cases this may make it difficult to argue for resources to be spent on safety.

Since Safety–I is defined as the 'freedom' from (unacceptable) risks, the phenomenology is really this 'freedom', i.e., the fact that nothing happens. However, as has been argued previously, Safety–I is in reality defined in terms of its opposite, namely the lack of safety. This means that we can say that a system is *unsafe* when something happens. We also tend to believe that a system is safe – or rather, *not unsafe* – when nothing happens, although this, strictly speaking, is not a valid logical conclusion. This 'reverse' definition creates the interesting practical question,

How to measure an increase in safety by counting how many fewer things go wrong?

The Aetiology of Safety–I

Since the phenomenology of Safety–I refers to the things that go wrong or can go wrong, adverse outcomes as well as adverse events, the aetiology of Safety–I must perforce be about the ways in which this can happen. In other words, the aetiology must be about the probable or possible causes and 'mechanisms' that produce the manifestations. The aetiology thus represents the accident models that are associated with Safety–I.

In the early days of safety thinking, corresponding to the era of simple linear causality and the age of technology, cf., Figure 2.1, accidents were seen as the culmination of a series of events or circumstances that would occur in a specific and recognisable order. Heinrich actually proposed a set of ten 'axioms of industrial safety', of which the first stated:

> The occurrence of an injury invariably results from a completed sequence of factors – the last one of these being the accident itself. The accident in turn is invariably caused or permitted directly by the unsafe act of a person and/or a mechanical or physical hazard.

And as is noted in Figure 3.3, accidents (the manifestations of Safety–I) could be prevented by finding and eliminating the possible causes.

In the following ages of Safety–I, the era of composite but still linear causality roughly covering the 1970–1990 period, accidents were seen as a result of a combination of active failures (unsafe acts) and latent conditions (hazards). As is noted in Figure 3.4, accidents could be prevented by strengthening barriers and defences. The generally held belief that accidents are the consequence of preceding causes is encapsulated in the pyramid model, as discussed above.

The aetiology of Safety–I thus includes assumptions about causality (cf., the *causality credo*), as well as the assumption that the results – the manifestations of Safety–I – can be explained by decomposition and by referring to the characteristics of the components, especially the ways in which they can fail or malfunction. The aetiology thus represents the purported mechanisms of (the lack of) safety. These can either be simple or composite linear developments, as illustrated by the Domino model and the Swiss cheese model respectively. Other, more complicated but still linear schemas of explanation can be found in the guise of specific methods or approaches, such as Tripod, AcciMap, or STAMP. According to Tripod, accidents happen through a combination of active and latent failures, such as dysfunctional barriers. AcciMap explains accidents in terms of an abstraction hierarchy of causes. And STAMP uses a model of socio-technical control comprising two hierarchical control structures as a basis for its explanations. The aetiology, however, also refers to something even more fundamental, namely the basic assumptions about the nature of how things happen.

Whenever something happens, it is assumed that there is a preceding cause. This principle was formulated by Leucippus of Miletus (c. 480–c. 420 BC), who is credited with the statement that 'Nothing happens in vain, but everything from reason and of necessity'. If this is so in general, then it is certainly also the case for situations where something goes wrong, i.e., incidents and accidents. Adverse outcomes are the result of adverse events, and adverse events have a cause – as expressed by Heinrich's first axiom of industrial safety. When human factors engineering and the study of man–machine systems became a serious business, roughly around the 1950s, systems were relatively easy to describe and understand and the causality assumption accordingly made good sense in most cases. It was generally possible to determine why things failed and to do something about it. But a large number of systems have by now become so difficult to describe and understand that it is practically impossible to unravel the cause–effect relationships that we traditionally assume exist. While the phenomenology remains the same, namely the things that go wrong, the aetiology is no longer unequivocal.

The Ontology of Safety–I

The relationship between the aetiology and the ontology is in many ways constrained, and it is impossible to say which is more important or which comes first (and 'first' itself is a concept loaded with assumptions about causality). Just as the aetiology is an account of how failures can lead to unwanted consequences – the 'mechanisms' by which failures and malfunctions (manifest or latent) can interact – the ontology is about the nature of failures, hence the true nature and the essential characteristics of safety. A closer look at the ontology – on that 'which is' – reveals that this involves three fundamental assumptions, which have been mentioned already, namely that systems are decomposable, that the functioning of the components can be described in bimodal terms, and that it is possible to determine the order in which events will develop in advance. We shall consider each of these in turn.

First Assumption: Systems are Decomposable

We have already encountered the issue of decomposability, above, as part of the overall process of deconstruction, and decomposition has been discussed relative to the concepts of reductionism. The Greek philosophers advocated not only the principle of causality but also the principle of decomposition, most famously in the theory of Democritus that everything is composed of 'atoms', which are physically indivisible. (Democritus was a student of Leucippus.) This has made it almost unavoidable to think of systems in terms of their components and how these components fit together in something that commonly is called the structure. The traditional definition of a system has therefore been with reference to its structure, i.e., in terms of the parts and how they are connected or put together. A simple version is to declare that a system is anything that consists of parts connected together. A slightly more elaborate one is that a system is 'a set of objects together with relationships between the objects and between their attributes'.

It makes good sense to assume that systems can be decomposed into their parts. We know that we can build systems (including aircraft, nuclear power plants, telephone networks and hospitals) by putting things together, and by carefully combining and organising components. We therefore also believe that this process can be reversed and that we can understand systems by decomposing them into meaningful constituents. We do have some success with decomposing technological systems to find the causes of accidents – the recent impasse of the Boeing 787 battery problems notwithstanding. We also assume that we can decompose 'soft systems' (organisations or socio-technical systems) into their constituents (departments, agents, roles, stakeholders). And we finally assume that the same can be done for tasks and for events, partly because of the seductive visual simplicity of a timeline or a hierarchical task description. But we are wrong in both cases; this goes both for the 'soft' systems and for what happens (events). In the case of organisations, we assume that we can decompose them into their constituents, for example, we see them represented in an organisational chart – never mind that we also know there is a crucial difference between the formal and the informal organisation. In the case of a workplace, we have experience with separating it into its constituents, such as in a human–machine system or in a distributed system (such as distributed decision-making). Indeed, the very notion of 'distributed' assumes that there are some things that can be distributed, each of which can be in different places but still able to function together in some essential manner. And in the case of things that happen, activities or events, we also have experience that we can break them into smaller parts – either as done by the time and motion studies of Scientific Management, or as done in various forms of task analysis and multilinear event sequence methods.

Second Assumption: Bimodality

When adverse outcomes have been traced back to their underlying causes, it is assumed that the 'components' that are part of the explanations either have functioned correctly or have failed. This principle of bimodality represents the commonly held assumption that things function until they fail. Whenever an individual component fails, such as a light bulb that burns out, the component will be discarded and replaced by a new (and usually identical) component. The same reasoning applies to composite systems although failures sometimes may be intermittent, especially if complicated logic (software) plays a part.

But even for composite – or non-trivial – systems, it is assumed that performance basically is bimodal: either the system works correctly (as designed) or it does not. The principle of bimodality means that the system and/or the system components can be described as being potentially in one of two different modes or states, either functioning or not functioning – not excluding the possibility of a zone of degraded operation in between. Systems are usually designed or engineered to provide a specific function and when that does not happen, for one reason or another, they are said to have failed or malfunctioned.

There is also an addendum to the second assumption, that has been mentioned in the discussion of the myths of Safety–I. This is the notion that there must be a kind of congruence or proportionality between causes and effects. If the effects are minor or trivial – a slight incident or a near miss – then we tacitly assume that the cause also is minor. Conversely, if the effects are major – such as a serious accident or disaster – then we expect that the causes somehow match that, or at least that they are not trivial. (A consequence of that is the assumption that the more serious an event is, the more we can learn from it; this is discussed further in Chapter 8.) In both cases the assumption about proportionality may clearly bias the search for causes, so that we look either for something trivial or something significant.

The Letter 'T'

The reader may well wonder about the reason for the heading of this section. The simple explanation is that the previous sentence started with the letter 'T' – as did this. But perhaps some further explanation is required.

Consider a manual typewriter, of the kind that I, and readers of an advanced age, used in former times. On a manual typewriter, if you hit the key marked 't', the type bar with the letter 't' would swing up and hit the ribbon, leaving a 't' printed as the next letter in the text you were writing. Hitting the key was the cause, and the printed letter was the effect. For a manual typewriter it was easy to understand how this happened, since there was a simple mechanical link between the key and the type bar. Even for an electrical typewriter, it was possible to understand how it worked, although there now were some hidden intermediate steps.

The development of more advanced typewriters made things a little less obvious. In the famous IBM Selectric typewriter, the type bars were replaced by a type ball, a small spherical element with the letters moulded into its surface. The typewriter used a system of latches, metal tapes and pulleys driven by an electric motor in order to rotate the ball into the correct position and then strike it against the ribbon and platen. (Other brands used rotating type wheels which were a little easier to understand.) While there still was a recognisable cause-and-effect relationship between hitting the key on the keyboard and having the corresponding letter appear on the paper, the understanding of what went on was no longer straightforward. There was now an electromechanical linkage that somehow produced the letter, and while it was probably still a simple sequence of causes and consequences, it was no longer possible to understand it directly or even to describe it or to reason by analogy.

Jumping ahead to the present day, typewriters are mostly to be found in museums. Instead we rely on computers or tablets of various types. On a computer with a keyboard, it is still possible to imagine some kind of neat 'mechanical' process. But on tablets where the input, in this case the letter 't', is either spoken or drawn on a touch sensitive surface, the one-to-one correspondence has disappeared. Indeed, the outcome may be a 't', a 'd', or any other letter that bears some resemblance (as a homonym or an isomorph of the letter 't').

When we consider the constituents that result from the decomposition, we assume that their functioning can be described using binary categories, such as true–false, functioning–malfunctioning, work–fail, on–off, correct–incorrect, etc. This kind of bimodal description is useful both for analyses of what has happened – and in particular when something has gone wrong – and for analyses of future situations, for example when we design a system or do a risk assessment. It is simple to comprehend and it fits nicely into the logical formalisms that we often rely on to grasp compound systems and events, to render them comprehensible when they challenge our natural capability. But it is often wrong.

The Ceteris Paribus *Principle*

The assumptions about decomposition and bimodality in combination lead to the conclusion that the elements or constituents can be described and/or analysed individually. This is, of course, a very convenient assumption to make, since it means that we can deal with the constituents one by one. (The principle is also applied to accident investigations, which usually are done for each accident separately.) To the extent that they do influence each other, this is assumed to happen in a linear manner, cf., below. In other words, the whole is just the sum of the parts, or the whole can be expressed and understood as a (linear) combination of the parts. This principle can be seen as analogous to the *ceteris paribus* assumption that rules empirical research. *Ceteris paribus* can be translated as 'with other things the same' or 'all other things being equal'. By invoking the *ceteris paribus* assumption, it becomes possible to focus on one element or function at a time, since all the others are 'equal', hence unchanged and with no effect.

In science, the *ceteris paribus* assumption is fundamental to most forms of scientific inquiry. In order to do controlled experiments, it is common practice to try to rule out factors which interfere with examining a specific causal relationship. Empirical science assumes that we can control all of the independent variables except the one that is under study, so that the effect of a *single* independent variable on the dependent variable – or variables – can be isolated. Economics, for instance, relies on the *ceteris paribus* assumption to simplify the formulation and description of economic outcomes. Clinical – and social – experiments do so as well. That risk assessments do so too is clear from the way in which calculations of failure probabilities are done in, for instance, fault trees. The *ceteris paribus* assumption also reigns supreme in safety management, organisational development, quality control, etc. When an accident investigation leads to a number of recommendations, it is blissfully assumed that each recommended action will achieve its purpose on its own, independently of other recommendations and independently of what goes on around it. In the social sciences, and in particular in industrial psychology and cognitive engineering, this is known as the substitution myth, the common assumption that artefacts are value neutral in the sense that their introduction into a system only has the intended and no unintended effects.

Third Assumption: Predictability

A further (classical) assumption is that the order or sequence of events is predetermined and fixed. This makes good sense when we consider the origin of Safety–I thinking, which referred to processes that were explicitly designed to produce a given outcome – whether the processes were technical ones such as a power plant, human activities such as in accomplishing a task in a factory or production line, or providing a service. Indeed, when we design new ways of doing something, either as a new task or activity or as an improvement of something that is being done already, we habitually and perhaps unavoidably think linearly in terms of causes and effects, and try to arrange conditions so that the process – or activity – proceeds as planned and intended. This makes the process easier to manage and control and, therefore, also makes it more efficient – at least as long as things go as planned without interruptions and disturbances. On the other hand, we know from everyday experience that it is sometimes – and not rarely – necessary to change the order in which things are done, it is necessary to improvise and to adjust performance. So while most of the work processes and functions that we use progress in a predetermined way – which is good because society as a whole depends on that – we also know that variability is ubiquitous as well as indispensable. While it is understandable that we make simplifying assumptions in the way we represent events in order to be able to analyse them (fault trees, event trees, Petri nets, networks, task analysis, etc.), it is also seriously misleading. For an event tree (such as the traditional THERP tree), for instance, a difference in the sequence of events – even if it is just the reversal of two steps – requires a separate analysis and a separate event tree. This is clearly a powerful motivation for employing strong assumptions that allow descriptions and analyses to be as simple as possible.

It is clearly convenient in many ways to analyse a process or a system by looking at the constituent functions or components individually. But the very fact of the decomposition means that it is necessary to put everything together again at the end of the analysis, in order to understand how the system as a whole will function or perform. Since the decomposition assumed linearity, the assembly or aggregation also does. In other words, when the (re-)combinations are made, they assume that the interactions or relationships between components can be described as trivial (causal, loosely coupled) and unidirectional or non-interacting. This again makes good sense for the many processes and systems that are designed to work in a certain way in order to deliver certain products or services. And it also behoves us to make sure that they function as they should – one reason being that we often depend on the outcomes. Yet we fool ourselves if we assume that this is the case for all systems. While that assumption might have been reasonable in the first decades of the twentieth century when (working) life was relatively simple, it is not generally reasonable today. There are, of course, still many systems and processes where making this assumption is justifiable from a pragmatic position, or at least does no irreparable harm. But we also see a growing number of cases – at first often spectacular accidents but with increasing frequency also complicated situations of suboptimal functioning that do not involve or lead to accidents – where a simple decomposition–aggregation approach is neither possible nor warranted. The number of these cases is, ironically, growing due to our inability fully to comprehend the systems we build. The inevitable consequence of that should be to abandon this and the other simplifying assumptions.

There is, however, yet another assumption, which somehow has attracted less notice but which is just as important. This is the assumption that the influence from context/conditions is limited and furthermore that it is quantifiable. We all have to acknowledge that work does not take place in a vacuum, although at the same time we strive to design working situations and workplaces as if they did (cf., the *ceteris paribus* assumption described above). There always is an influence from the environment or the context, and in order to make a reasonable analysis we somehow have to take that into account. The problem has been recognised by many scientific disciplines, sometimes willingly and sometimes reluctantly. Decision theory, for instance, held out for most of its history (which goes back at least to Blaise Pascal, 1623–1662), but lately succumbed in the form of naturalistic decision theory, which was developed to describe how people actually make decisions in demanding situations. Economics tries but only in a half-hearted manner, and still maintains the assumption about market rationality. Behavioural sciences, and cognitive psychology, in particular, had to give up context-free cognition – or information processing – in lieu of 'situated cognition' and 'cognition in the wild'. Accident investigations obviously cannot afford to disregard the influence of the context, although it often leads to the conclusion that the accident was due to the very special conditions and that the system (or people) would have worked perfectly if such conditions could have been avoided – which often makes itself known in the recommendations. Risk assessment, particularly in the quantitative versions such as probabilistic risk assessment (PRA), stubbornly stick to the independence assumption. This can be seen both by the way that functions are described and analysed independently (cf., above), and by the way that the focus is still on the (failure) probability on a function-by-function basis. So while it is recognised that there is an influence from the conditions or the environment, it is represented as a set of performance-shaping factors that modifies the failure probability. The PSFs themselves are either elements or can be decomposed, and their combined influenced can be obtained by simple arithmetic operations – in other words on the basis of the assumptions of linearity and independence.

Ontology Summary

The point of all this is to argue that as the world – for instance, represented by the lowly typewriter – becomes more complicated, it is no longer reasonable to assume that we can understand the causal relationships between actions and outcomes, or even that they can be described in causal terms. While the psychological (and social?) need to explain everything in a simple way is undeniable, it may no longer be possible to satisfy that need and at the same time maintain an acceptable degree of realism and practicality. It may well be that outcomes appear despite any obvious actions or that the relationship between the action (activity) and the outcome is opaque and difficult to understand.

Even in the case of the typewriter we may, taking some liberties, talk about the phenomenology, the aetiology and the ontology (though the latter perhaps is going a bit too far). In a growing number of cases we must stop taking for granted that we can find the cause when something has happened – or even that we can come up with a plausible explanation. This is so regardless of whether the outcome is wanted or unwanted, although normally we are more worried about it in the latter case.

To put it simply: the ontology of Safety–I cannot be sustained. Or rather, Safety–I thinking is no longer universally applicable. We must keep in mind that even if we limit the focus to traditional safety concerns, this way of thinking was developed almost a century ago. The Domino model was described in a book published in 1931, but the experiences that led to the ideas and theories described in the book were from the preceding decades. The thinking that was relevant for the work environments at the beginning of the twentieth century is unlikely to be relevant today when socio-technical systems are not decomposable, bimodal, or predictable.

Comments on Chapter 5

The originator of deconstruction is the French philosopher Jacques Derrida. Deconstruction is presented in Derrida, J. (1998), *Of Grammatology*, Baltimore, MD: Johns Hopkins University Press.

The UK National Health Service definition can be found on the website http://www.npsa.nhs.uk/nrls/reporting/what-is-a-patient-safety-incident/. While the trend to include potentially harmful events is understandable, it is also rather unfortunate because the reporting of such events depends on how people have judged a situation rather than on what actually happened.

On the TRIPOD website (http://www.energypublishing. org/tripod), the method is described as 'a theory (sic!) for understanding incidents and accidents', in particular to 'allow the root organisational causes and deficiencies to be uncovered and addressed'. TRIPOD can be seen as an elaboration of the ideas in the Swiss cheese model. The AcciMap is a method to develop a 'map of an accident' by means of the abstraction hierarchy. The original description can be found in Rasmussen,

J. and Svedung, I. (2000), *Proactive Risk Management in a Dynamic Society*, Karlstad, Sweden: Swedish Rescue Services Agency. The Systems–Theoretic Accident Model and Processes (STAMP) approach has been developed by Nancy Leveson and described in several papers. It is a composite linear analysis method based on a view of systems as being hierarchically structured. The properties that emerge from a set of components at one level of hierarchy are controlled by constraining the degrees of freedom of those components, thereby limiting system behaviour to the safe changes and adaptations implied by the constraints. All three methods directly or indirectly reflect ideas developed by Jens Rasmussen during his tenure at the Risø National Laboratories in Denmark.

There are many different definitions of what a system is, but they are generally in good agreement. The definition used in this chapter can be found in Hall, A.D. and Fagen, R.E. (1968), Definition of system, in W. Buckley (ed.), *Modern Systems Research for the Behavioural Scientist*, Chicago, IL: Aldine Publishing Company.

The substitution myth is the common assumption that artefacts are value neutral, so that their introduction into a system has only intended and no unintended effects. The basis for this myth is the concept of interchangeability, which has been successfully used in the production industry. It is not unreasonable to assume that substitutability is possible for technical systems where the parts are not interacting and where there is no appreciable tear and wear. But it is not a reasonable assumption to make for socio-technical systems. A good discussion of the substitution myth is given in a chapter entitled 'Automation surprises', by Sarter, N.B., Woods, D.D. and Billings, C.E. (1997), which appears in Salvendy, G. (ed.), *Handbook of Human Factors & Ergonomics*, second edition, New York: Wiley.

Chapter 6
The Need to Change

Tempora mutantur ...

Chapter 5 concludes that the ontology of Safety–I has become obsolete because it has stayed the same while the world has changed. The changes that have happened can, for instance, be seen by comparing a work setting from the early 1970s with a work setting about 40 years later. Figure 6.1 shows how air traffic controllers (ATCs) were trained in 1970 and in 2013; the training environments can reasonably be assumed to represent the essential characteristics of the actual working environments. Other domains would show similar differences.

In the 1970s environment, there were no computers present in the workplace and, although computing was used in the background, the level of automation and computerised support was low. (The same goes for the pilots who were guided by ATCs.) The training environment shows that the volume of traffic was low, which also means that the pace of work was lower than today. And the focus was on what happened at a specific airport or in a specific sector, with limited coordination with other sectors. The contrast with the work environment of 2013 is stark. Work is not only supported by multiple computer systems – visible as well as invisible – but quite obviously also takes place by means of computers; traffic volumes are higher – roughly three times as high – leading to higher work demands and workload. The more complicated traffic situations make a tighter integration with neighbouring sectors necessary, in addition to the changing geography of sectors. Similar changes can be found offshore, in the production industries, in health care, in financial systems and even in a non-work situation such as driving a modern car through city traffic.

Figure 6.1   Changed training environments for air traffic
controllers

Source: With permission, DFS Deutsche Flugsicherung GmbH.

The Rate of Inventions

These changes can be seen as due to the mutual effect of two factors, one being human inventiveness and the other being the constant striving to increase our mastery of the world around us – today herostratically known as the principle of faster, better, cheaper. Human inventiveness has, however, never been steady or stable, but has varied through history. The British historian of science Samuel Lilley characterised the rate of change in terms of the relative rate of inventions, defined as the percentage increase in man's mechanical equipment occurring on average in one year. From a grand historical perspective stretching back to Neolithic times this rate peaked on several occasions (after the introduction of agriculture, in the great days of Athenian culture, at the end of the Hellenistic expansion and during the Renaissance), each time only to fall back to a lower level. Around the beginning of the eighteenth century, corresponding to the second industrial revolution, the rate started an upward climb that has so far seemed to have no ceiling.

Moore's Law

A contemporary expression of the rate of invention is Moore's Law, which refers to the observation that the number of transistors on integrated circuits has doubled approximately every two years. (In some version of the law the period is 18 months.) This growth rate seems to have been sustained from the initial formulation of the law in 1965 until today – although to some extent it may have become a self-fulfilling prophecy for chip designers.

A graphical rendering of Moore's Law shows the consequences of doubling every second year (Figure 6.2). The starting point is 2,300 transistors per central processing unit (CPU) in 1971. As of 2012, the highest transistor count in a commercially available CPU was over 2.5 billion transistors, while the world record for a field-programmable gate array (FPGA) was 6.8 billion transistors. No one has counted the number of gadgets and devices that have been created from this, but it is surely not inconceivable that the relative invention rate has passed the value of 1.0 by now.

Regardless of whether Moore's law is a precise formulation, and even if the period of doubling is three years instead of 18

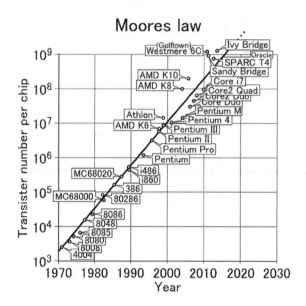

Figure 6.2   Moore's law

Source: Wikimedia.

months, the law represents the fact that the capabilities of electronic devices and, therefore, also of work equipment, grows at an exponential rate – at the same time as the devices become cheaper and smaller. This increases the capacity of processors, sensors and networks leading to more complicated and more tightly coupled machines and thereby to a changed socio-technical habitat.

The second factor accounting for the increased rate of inventions is the constant human striving to master the world, not so much in the conventional sense of conquering it, but in the sense of controlling it, so that the number of unexpected situations is acceptably small. This goes hand in hand with an insatiable desire to do things with greater speed, with greater precision and with less effort. One formulation of this is the Law of Stretched Systems, which states, 'Every system is continuously stretched to operate at its capacity'. Combined with the continued growth in the relative rate of inventions, the result is a continued stretching of system capabilities. The term 'stretching' implies that the expansion is going for the maximum ability, i.e., stretching to the limits. The sad consequence of that is that we develop or expand functionality incrementally, without being able to fully comprehend the consequences of what we do.

The Lost Equilibrium

The irony of the combined effect of the two factors is that it creates an unstable situation, in the sense that we develop systems that we are unable to control – sometimes even deliberately so, as in the case of dynamically unstable fighter aircraft. This should, all things being equal, lead to a state of equilibrium, where the development of new equipment would be kept in check by our ability to use and control it. That equilibrium has unfortunately been defeated by a human hubris, nourished by a cocktail of ingenuity and optimism, which uses the power of technology to compensate for our inability to control what we build. The main mode of compensation is automation. The belief in automation has remained strong for the last 50–60 years, even though it violates the substitution myth. Taken together, these developments create a self-reinforcing cycle, as shown in Figure 6.3.

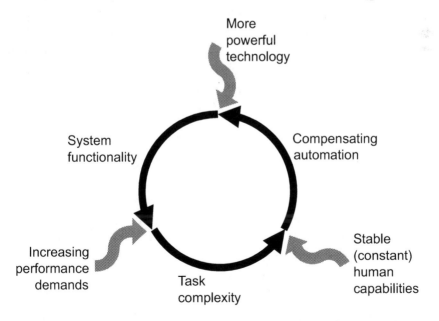

Figure 6.3   **Self-reinforcing cycle of technological innovation**

It gives little comfort that Norbert Wiener characterised this situation as far back as 1954 by noting, '[W]e have modified our environment so radically that we must now modify ourselves in order to exist in this new environment'. (And today's infatuation with solutionism only makes things worse.) Technically, the situation may be characterised by noting that system phenomena are emergent, whereas our thinking still is based on pedestrian causal logic that limits us to resultant outcomes. System developments occur in parallel, while we are limited to think sequentially – at least when we have to rely on conscious or intentional thinking (as in designing something or making a safety assessment). Systems have therefore become intractable (cf., below), which means that we no longer understand what will happen, how it will happen, or why it will happen.

A conspicuous consequence of this is the way that computing machinery and IT devices have invaded everyday life and changed it beyond recognition. (This has also made the study of human–machine interaction, or human–computer interaction, a major field, scientifically and as an industrial practice.) However, the manifest uses of computing machinery with its mixture of benefits and problems (among the latter the dreaded case of automation surprises), pales in relation to how these developments have changed the socio-technical habitat and the nature of work. This development is only visible in hindsight, one reason being that it is often several layers removed from the people who use it – with the consequence that they are not really aware of it. While the importance of issues such as interface design, usability, and 'intelligent' devices is steadily growing, machine–machine communication goes unnoticed by many. However market predictions recognise that this is a growing area and also, rightly, a growing concern. To illustrate the scope of this development, industry pundits and researchers estimate that there will be more than six billion wireless subscribers using smartphones by 2020. This may in fact be too conservative, since according to a United Nations (UN) agency report there will be more mobile subscriptions than people in the world by the end of 2014. And communications companies such as Ericsson predict that by 2020 there will also be more than 50 billion intelligent machines, in large industrial installations, in small devices, in homes, in vehicles and probably in every possible and impossible place as well. More than 50 billion intelligent machines will inevitably bring about such a radical modification of our socio-technical habitat that we should start to think about it now, lest we want to find ourselves in the unenviable position of the sorcerer's apprentice. (As an aside, the machines will mostly communicate with themselves. As an illustration of that, BBC News on 12 December 2013 reported that 61.5 per cent of all web traffic was generated by bots, an increase of 21 per cent since the year before.)

The bottom line is that the traditional ways of thinking about safety and other issues (decomposition, bimodality, predictability) are totally inadequate as a basis for designing such systems, as a basis for operating them and as a basis for managing their performance – and especially for understanding what happens when something goes wrong. Since work today takes place in a socio-technical habitat that is both complicated and tightly coupled, our models and methods must necessarily reflect that fact. This is so regardless of whether we look at safety, productivity, quality, resilience and so on.

The New Boundaries

Today's challenge is to develop and manage systems that are larger and more complicated than the systems of yesteryear. The differences can be described in several ways, focusing on the structure – or the nature – of the systems, or on their scope or extent. Since the concern is how to manage systems, the scope refers to the functional rather than the physical characteristics. One aspect of that is the time span that system management needs to cover or consider. Another is the number of organisational layers that need to be considered. And the third is the extension of the system, in terms of whether it can be considered by itself as an independent system, or whether there are dependencies to other systems that must be taken into account. The latter can also be seen as the question of where to put the boundary of the system. Without going into a lengthy discussion of the nature of what a system is and what a system's environment is, it can simply be noted that the delineation of either is relative. Peter

Checkland, who for many years has been one of the leading figures in systems thinking, wrote about the boundary between a system and its environment as:

> the area within which the decision-taking process of the system has power to make things happen, or prevent them from happening. More generally, a boundary is a distinction made by an observer which marks the difference between an entity he takes to be a system and its environment.

Time Span or Temporal Coverage

The natural focus for safety management and, indeed, for any kind of process management, is what the system does or what it produces, since that is its raison d'être. The time span of interest is therefore the typical duration of an activity or operation. For a cruise liner it is the duration of a cruise. For a surgeon it is the duration of an operation. For a trader it is the duration of a trade session (since a single trade by today has become extremely brief and, moreover, taken over by computers). For a power plant operator it is the duration of a shift, but for a power plant owner it is the time between planned outages.

The typical durations can thus vary considerably and be counted in seconds, in hours, in days, in weeks, or in months. One consequence of the developments described above is that it has become necessary to extend the time span that an analysis usually considers, to include what happens earlier and what happens later in the system's life. What happens earlier may affect the ability to operate the system, hence its ability to function as intended. This includes, of course, the preparations or set up, establishing the necessary preconditions, making sure that tools, data, resources, etc. are in place. But it also includes issues and conditions that lie further back in time, for instance design decisions (an example being the use of lithium batteries in the Boeing Dreamliner). Similarly, it is also necessary to consider what happens later in the system's life, for instance in terms of long-term effects on the environment. Some concerns may also be needed in relation to the possible dismantling of the system, the decommissioning or retirement and all that follows from that. (This is a well-known problem for nuclear power plants, but applies also to many other industries in terms of the ecological 'footprint' they make on the local or global environment.) Decisions that are made during an operation today may have consequences far down the line and the management of the system (in terms of safety, quality, or productivity) has to take that into account.

Number of Organisational Layers or Levels

Just as it has become necessary to consider what happens earlier and later in the system's life, eventually spanning the whole time span from creation to dismantling, it has also become necessary to consider what happens at other system levels or layers relative to the actual operations. This has to some extent been acknowledged for decades, as is illustrated by the introduction in the 1990s of the notions of the sharp end and blunt end, mentioned previously.

The sharp end denotes the people at work at the time and place where something happens, while the blunt end denotes people 'removed in time and space' from actual operations. The terms were introduced to improve the explanation of how accidents happened, but their use is perfectly legitimate to enable understanding of how work takes place. However, the terms, need to be enriched. What happens during actual day-to-day operations – the sharp end – is clearly affected by what has happened in organisational strata 'above' the operations. This may refer to the detailed planning of work, work norms and quotas, quality and quantity, types of work products, maintenance of equipment (and of human competence), organisational culture, etc. Indeed, the upwards extension of interest may easily stretch to the very top of an organisation and beyond that, to company boards, shareholders, regulators, etc. In the same way, the ability to carry out day-to-day operations is clearly affected by the strata 'below' (meaning with less scope or ability to decide), such as actual maintenance activities, providing resources, carrying out 'menial' tasks (that are assumed simply to be done, such as cleaning and renewal of basic resources), making sure that spare parts and reserves are in place, etc. In both cases, this is something that happens before the actual operations, but which still is directly related to them.

Extension or Horizontal Couplings – the Boundary

The third aspect of the scope, the extension, is of a different kind since it refers to the relationships between the system and its environment. The environment is, however, neither a homogeneous nor a passive entity. The environment itself consists of other systems. (The terms 'system' and 'environment' are thus relative, depending on the chosen focus, similar to the notions of figure and ground in Gestalt psychology or visual perception.)

The relationships are temporal in the sense that something that has happened before may affect the system's operations, just as the operations may affect something that happens later. Things that have happened before and which affect the system are called *upstream* events, while those that happen later and which may be affected by the system's operations are called *downstream* events. (In a sense, the blunt end is always upstream to the sharp end, but there is more in the sharp end–blunt end distinction than the temporal relationship.)

One illustration of extended horizontal couplings was provided by the earthquake that hit Japan on 11 March 2011. While the interest for many reasons has focused on the consequences that the earthquake and the following tsunami had for the Fukushima Daiichi nuclear power plant, there were other consequences as well. Japan's leading automaker, Toyota, has a large presence in northern Japan, and was therefore most affected by the quake. The earthquake did not just damage a number of Toyota's own production sites, but also effectively disrupted the supply chain. Modern manufacturing methods require a steady flow of parts and materials from upstream sources to arrive just in time and any disruption of this can be fatal. This dependence makes it necessary to extend the boundary of the system to include events that happen upstream, in order to insure against disruptions as far as possible. It may similarly be necessary to extend to boundary to include what happens downstream, in order to prevent unwanted consequences of the operations.

On the whole, it becomes necessary to think about functional dependences between the local system and other systems, and to take steps to understand the dependences and vulnerabilities. In other words, it becomes necessary to extend the scope of the

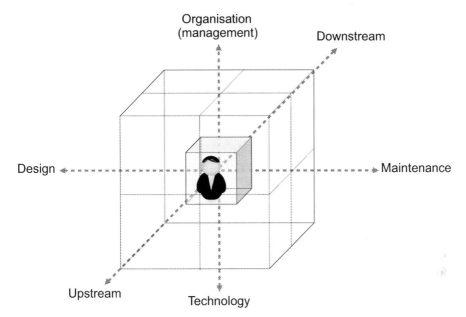

Figure 6.4   The enlargement of the safety focus

system from the local, represented by the small cube in Figure 6.4, to the global, represented by the large cube.

In terms of defining the boundary between a system and its environment – following Checkland's recommendation – and remembering that the environment can be described as the net effects of the performances of a number of other systems, Table 6.1 provides some practical hints.

Table 6.1   A pragmatic definition of a system's boundary

	Other systems whose functions are important for the management of the target system	Other systems whose functions are of no consequence for the management of the target system
Other systems that can be effectively controlled by the target system	1. Other systems should be included in the target system (inside the boundary)	2. Other systems may be included in the target system (inside the boundary)
Other systems that cannot be effectively controlled by the target system	3. Other systems cannot be included in the target system (outside the boundary)	4. Other systems should be excluded from the target system (outside the boundary)

In consequence of these developments safety concerns must today address systems that are larger and more complicated than the systems of yesteryear. Because there are many more details to consider, because some modes of operation may be incompletely known, because of tight couplings between functions, and because systems may change faster than they can be described, the net result is that many systems are underspecified or intractable. For these systems it is clearly not possible to prescribe tasks and actions in every detail and we must therefore relinquish the notion that Work-As-Imagined will correspond to Work-As-Done. On the contrary, for work to succeed it is necessary that performance is variable or flexible rather than rigid. In fact, the less completely a work system can be described, the more performance variability is needed.

Tractable and Intractable Systems

In order for a system to be controllable, it is necessary to know what goes on 'inside' it and to have a sufficiently clear description or specification of the system and its functions. The same requirements must be met in order for a system to be analysed, so that its risks can be assessed. That this must be so is obvious if we consider the opposite. If we do not have a clear description or specification of a system, and/or if we do not know what goes on 'inside' it, then it is clearly impossible to control it effectively, as well as to make a risk assessment. We can capture these qualities by making a distinction between tractable and intractable systems, cf., Table 6.2 opposite. A system is tractable if the principles of its functioning are known, if descriptions of it are simple and with few details and, most importantly, if it does not change while it is being described. An example could be an assembly line or a suburban railway. Conversely, a system is intractable if the principles of its functioning are only partly known (or, in extreme cases, completely unknown), if descriptions of it are elaborate with many details and if systems change before descriptions can be completed. An example could be emergency management after a natural disaster or, *sans comparaison*, financial markets.

Table 6.2   Tractable and intractable systems

	Tractable system	Intractable system
Number of details	Descriptions are simple with few details	Descriptions are elaborate with many details
Comprehensibility	Principles of functioning are known	Principles of functioning are partly unknown
Stability	System does not change while being described	System changes before description is completed
Relationship to other systems	Independence	Interdependence
Controllability	High, easy to control	Low, difficult to control
Metaphor	Clockwork	Teamwork

It follows directly from the definition of tractability that an intractable system also is underspecified. The consequences are that the predictability is limited and that it is impossible precisely to prescribe what should be done. Underspecification is, of course, only an issue for the human and organisational parts of the system. For the technical parts, in so far as they work by themselves, complete specification is a necessity for their functioning. An engine or a machine can only function if there is a description of how every component works and how they fit together. It is in this sense that the performance of a technical system *results* from the parts. Technological systems can function autonomously as long as their environment is completely specified and preferably constant, in the sense that there is no unexpected variability. But this need of a complete technical specification creates a dilemma for socio-technical systems. For such systems the environment cannot be specified completely and it is certainly not constant. In order for the technology to keep working, humans (and organisations) must function as a buffer between subsystems and between the system and its environment, as something that absorbs excessive variability when there is too much of it and provides variability when there is too little. The problem can in some cases be solved by decoupling parts of the system, or by decomposing it. But for an increasing number of systems this solution is not possible.

The Reasons Why Things Work Revisited

Altogether, this means that performance variability is inevitable, both on the level of the individual and on the level of the social group and organisation. At the same time, performance variability is also needed, as is argued in the previous section. This changes the role of the people, of the human factor, radically. During the history of system safety, the human factor has been considered first as a bottleneck that hindered the full use of the technological potential and then as a liability or a threat that not only limited performance but also was a source of risk and failure ('human error'). But by recognising the necessity and value of performance variability, the human factor becomes an asset and a sine qua non for system safety. This change in perception has not come about abruptly, however, but has been slowly growing for the last 15 years or so. To be consistent with this role, accident analysis and risk assessment must acknowledge the following:

- Systems are not flawless and people must learn to identify and overcome design flaws and functional glitches.
- People are able to recognise the actual demands and can adjust their performance accordingly.
- When procedures must be applied, people can interpret and apply them to match the conditions.
- People can detect and correct when something goes wrong or when it is about to go wrong, and hence intervene before the situation seriously worsens.

Taken together these premises describe work as actually done rather than work as imagined. They describe systems that are real rather than ideal. Such systems are usually highly reliable, but are so because people are flexible and able to adjust rather than because the systems have been perfectly thought out and designed. Under such assumptions, humans are no longer a liability and performance variability is not a threat. On the contrary, the variability of everyday performance is necessary for the system to function and is the source of successes as well as of failures. Because successes and failures both depend on performance variability, failures cannot be prevented by eliminating it; in other words, safety cannot be managed by imposing constraints on how work is done. The solution is instead to identify the situations where the variability of everyday performance may combine to create unwanted effects and continuously monitor how the system functions in order to intervene and dampen performance variability when it threatens to get out of control. At the same time, we should also keep an eye on situations where variability may have useful effects, and learn how to manage and reinforce that.

Work-As-Imagined comes from the tradition established by Scientific Management Theory of decomposing tasks and activities as a starting point for improving work efficiency. The undeniable accomplishments of this technique provided the theoretical and practical foundation for the notion that Work-As-Imagined was a necessary and sufficient basis for safe and effective work. Adverse outcomes could therefore be understood by considering the preceding events as components, and finding those that had failed. Safety could similarly be improved by carefully planning work in combination with detailed instructions and training. This is recognisable in the widespread belief in the efficacy of procedures and the emphasis on procedure compliance. In short, safety could be achieved by ensuring that Work-As-Done was identical to Work-As-Imagined.

Many organisations seem willing to assume that complicated tasks can be successfully mapped onto a set of simple, Tayloristic, canonical steps that can be followed without need of understanding or insight (and thus without need of significant investment in training or skilled technicians). But except for extremely simplified work situations this is not a realistic assumption to make. Instead, the work that is carried out – appropriately called Work-As-Done – is neither simple nor predictable in the manner of Work-As-Imagined. As is described above, people always have to adjust work to the actual conditions, which on the whole differ from what was expected – and many times significantly so. This is the performance adjustments or the performance variability that is at the core of Safety–II, which will be presented in greater detail in the next chapter. The distinction can be found in French psychology of work from the 1950s, where the terms were *tâche* (for task, or Work-As-Imagined) and *activité* (for activity, or

Work-As-Done) respectively. The distinction has been taken up by resilience engineering and been very useful there.

Work-As-Imagined tends to see the action in terms of the task alone and cannot see the way in which the process of carrying out the task is actually shaped by the constantly changing conditions of work and the world. An analogy, although inadequate, is the difference between a journey as actually carried out on the ground and as seen on a map. The latter inevitably smoothes over the myriad decisions made with regard to imprecise information and changing conditions in the road environment, in the traffic and in the drivers. Although the map is very useful, it provides little insight into how ad hoc decisions presented by changing conditions can be resolved (and, of course, each resolved decision changes the conditions once more). As a journey becomes more complicated, the map increasingly conceals what is actually needed to make the journey.

By way of conclusion, the more complicated environments that are characteristic of today means that Work-As-Done is significantly different from Work-As-Imagined. Since Work-As-Done by definition reflects the reality that people have to deal with, the unavoidable conclusion is that our notions about Work-As-Imagined are inadequate if not directly wrong. This constitutes a serious but healthy challenge to the models and methods that comprise the mainstream of safety engineering, human factors and ergonomics. We must be willing to meet that challenge head-on. Otherwise we may inadvertently create the challenges of the future by trying to solve the problems of the present with the models, theories and methods of the past.

Comments on Chapter 6

When Dan Golding became administrator of NASA in 1992, he introduced an approach called 'faster, better, cheaper' to space missions. The 'faster, better, cheaper' principle has been the subject of much dispute and is generally seen as being unrealistic. It has been discussed in Woods, D.D. et al. (2010). *Behind Human Error*, Farnham: Ashgate.

In 1948, Sam Lilley presented an original and interesting analysis of the development and use of technology in *Man, Machines and*

History, London: Cobbett Press. To give a feeling for what the relative rate of invention means, the value at the Renaissance was estimated to be .11, or roughly a 10 per cent contribution in the form of new technologies year by year. The corresponding value around 1900 was estimated to be about .6; it is anybody's guess what it is today. Moore's Law is named after Gordon E. Moore, who in 1965 worked at Fairchild Semiconductors but in 1968 left and became one of the founders of Intel. It was described in Moore, G.E. (1965), Cramming mode components onto integrated circuits, *Electronics Magazine*, 114–17.

The Law of Stretched Systems can be expressed in several ways, for instance as meaning that 'every system is stretched to operate at its capacity; as soon as there is some improvement, for example in the form of new technology, it will be exploited to achieve a new intensity and tempo of activity'. It has been discussed in Woods, D.D. and Cook, R.I. (2002), Nine steps to move forward from error, *Cognition, Technology & Work*, 4, 137–44.

Norbert Wiener's quotation is taken from Wiener, N. (1954), *The Human Use of Human Beings*, Boston, MA: Houghton Mifflin Co. Despite being written 60 years ago, this book is still well worth reading. Some years later Wiener wrote about 'gadget worshippers, who regard with impatience the limitations of mankind, and in particular the limitation consisting in man's undependability and unpredictability' (Wiener, N. (1964), *God & Golem, Inc.: A Comment on Certain Points Where Cybernetics Impinges on Religion*, Cambridge, MA: MIT Press.) In other words, about people who think that there is a technological solution ('gadget') to every problem. This is very similar to the contemporary idea of 'solutionism', which has been defined as 'an intellectual pathology that recognizes problems as problems based on just one criterion: whether they are "solvable" with a nice and clean technological solution at our disposal' by E. Morozov, in The perils of perfection, *The New York Times*, 2 March 2013. Solutionism has two practical consequences. One, that problems are attacked and solved one by one, as if they could be dealt with in isolation. The other, that the preferred solution is technological rather than socio-technical, probably because non-technical solutions rarely are 'nice' and 'clean'.

The modern recognition of the importance of distinguishing between Work-as-Imagined and Work-as-Done should be attributed to Leplat, J. and Hoc, J.M. (1983), Tache et activite dans l'analyse psychologique des situation, *Le Travail humain*, 3(1), 49–63. An earlier reference in the French ergonomics literature is Ombredane, A. and Faverge, J.M. (1955), *L'analyse du travail*, Paris: Presses Universitaires de France.

Finally, Peter Checkland's writings about systems theory and systems thinking have been collected in Checkland, P. (1999), *Systems Thinking, Systems Practice*, New York: Wiley.

Chapter 7
The Construction of Safety–II

The purpose of the deconstruction of Safety–I was to find out whether the assumptions are still valid – which in turn means whether the perspective offered by Safety–I remains valid. The deconstruction of Safety–I showed that the phenomenology refers to adverse outcomes, to accidents, incidents and the like; that the aetiology assumes that adverse outcomes can be explained in terms of cause–effect relationships, either simple or composite; and that the ontology assumes that fundamentally something can either function or malfunction.

Following on the deconstruction, Chapter 6 argues that work environments have changed so dramatically during the last two to three decades that the assumptions of Safety–I no longer are valid. Chapter 6 describes more precisely why and how this has happened and explains the consequences of this new reality. The purpose of Chapter 7 is therefore to *construct* a perspective on safety that makes sense vis-à-vis the world that we have to deal with now and in the near future, by reversing the process of deconstruction described in Chapter 5. This new perspective will, not surprisingly, be called 'Safety–II'. The construction will reverse the process of deconstruction and therefore begin with the ontology of Safety–II, then go on to the aetiology of Safety–II and end with the phenomenology of Safety–II.

The Ontology of Safety–II

As is argued above, an increasing number of work situations are becoming intractable, despite our best intentions to the contrary. One of the reasons for this is, ironically, our limited ability to understand completely what we do, in the sense of being able to anticipate the consequences of design changes and other types of well-meant interventions to improve safety, quality, productivity, etc. The intended consequences are more often than not a product of wishful thinking, based on an oversimplified understanding of how the world works. (Examples of that from, for example, heath care, public transportation and the worlds of finance or domestic and international politics are abundant.) In the industrial domains it is generally assumed that everything – including people – works as imagined, as we think it works. The unintended consequences or side effects are often missed due to a lack of (requisite) imagination or to what the American sociologist Robert Merton formulated, as long ago as 1936, as 'the law of unintended consequences'. (Another explanation is that people in their eagerness to get things done make the trade-offs between efficiency and thoroughness that the ETTO principle describes.) Indeed, the difficulties in comprehending what is going on is one of the dimensions of the tractability–intractability characterisation. This problem was addressed 30 years ago when the British psychologist Lisanne Bainbridge, in a discussion of automation, pointed out that 'the designer who tries to eliminate the operator still leaves the operator to do the tasks which the designer cannot think how to automate'. This argument is not only valid for automation design but applies to work specification and workplace design in general.

We can only specify the work in detail for situations that we understand completely, but there are very few of those. Because we cannot specify work in detail for the rest, human performance variability is indispensable. There are, in fact, only very few situations where performance adjustments are not required, even if every effort is made to establish an isolated and fixed work environment, for instance sterile rooms in neonatal intensive care units, or ultra-clean rooms in a unit where electronic chips are manufacturered. Even in a strictly regulated military operation such as the changing of the guard in front of a royal palace, the soldiers will have to adjust to wind, weather and intrepid tourists. The more complicated or less tractable a work situation, the greater the uncertainty about details and the greater the need for performance adjustments. (That is, more will be left to human experience and competence, and less to the capabilities of technological artefacts.)

The ontology of Safety–II is consistent with the fact that many socio-technical systems have become so complicated that work situations are always underspecified, hence partially unpredictable. Because most socio-technical systems are intractable, work conditions will nearly always differ from what has been specified or prescribed. This means that little, if anything, can be done unless work – tasks and tools – are adjusted so that they correspond to the situation. Performance variability is not only normal and necessary but also indispensable. The adjustments are made by people individually and collectively, as well as by the organisation itself. Everyone, from bottom to top, must adjust what they do to meet existing conditions (resources and requirements). Because the resources of work (time, information, materials, equipment, the presence and availability of other people) are finite, such adjustments will always be approximate rather than perfect. The approximation means that there is inevitably a small discrepancy between what ideally should have been done, or the 'perfect' adjustment, and what is actually done. Yet the discrepancy is usually so small that it has no negative consequences or can be compensated for by downstream adjustments. This is so regardless of whether the discrepancy is found in one's own work or in the work of others.

It is somewhat ironic that performance adjustments must be approximate because of the very conditions that make them necessary. In other words, if there were sufficient time, information, etc., in a work situation, then performance adjustments would not be required. (This does not rule out that they might happen for other reasons, such as boredom or the human creative spirit.) But as it is, they are required because the 'resources' are not completely adequate, which in turn means that the adjustments cannot be perfect or complete but must remain approximate.

The ontology of Safety–II is thus that human performance, individually or collectively, always is variable. This means that it is neither possible nor meaningful to characterise components in terms of whether they have worked or have failed, or whether the functioning is correct or incorrect. The bimodality principle of Safety–I is therefore obsolete.

Performance variability should, however, not be interpreted negatively, as in 'performance deviations', 'violations', and 'non-compliance'. On the contrary, the ability to make performance adjustments is essential for Work-As-Done. Without that anything but the most trivial activity would be impossible. And we would certainly not have the complicated nest or network of activities necessary to sustain the performance of contemporary societies – be they industrialised or developing nations.

The Aetiology of Safety–II

The aetiology is the description of the ways in which things happen, or rather the assumptions about how things happen. It is the description of the 'mechanisms', the simplified explanations, that can be used to make sense of what is observed, of what happens, hence also to manage it. The aetiology is the way of explaining the phenomenology in terms of the ontology.

Whenever something happens, in a system or in the world at large, whenever there is an unexpected event or an unexpected outcome, an explanation is sought. In most cases the explanation relies on the general understanding of how systems function, which means that it includes the principles of decomposition and causality. In such cases the outcome is said to be a result of the 'inner' workings of the system and is therefore technically called 'resultant'. But since the ontology of Safety–II is based on performance adjustments and performance variability rather than the bimodality principle, the aetiology cannot simply be the principle of causality and linear propagation of causes and effects that is used in Safety–I.

Explanations of the majority of adverse events can in practice still be expressed in terms of a breakdown or malfunctioning of components or normal system functions, at least as a trade-off between thoroughness and efficiency. Most adverse events have minor consequences and are therefore not subject to extensive analyses, but treated more lightly as near misses or incidents (although not in the oversimplified manner of the pyramid model discussed earlier). The reason why there are far more events of the less serious kind (such as runway incursions or bed ulcers) than of the more serious kind (such as 'airproxes' (near misses by aircraft in flight) or 'wrong-side' surgery) is simply that most systems are protected against serious events by means of an extensive system of barriers and defences. For serious adverse events that may occur regularly, more thorough precautions are taken and more attention is paid.

There is, however, a growing number of cases where such a trade-off is not acceptable, and where it is impossible to explain what happens by means of known processes or developments. It is still possible to provide an explanation of what happened, but it is a different kind of explanation. In such cases the outcome is said to be 'emergent' ('rising out of' or 'coming into' existence) rather than resultant ('being a consequence of'). The meaning of 'emergence' is not that something happens 'magically', but simply that it happens in such a way that it cannot be explained using the principles of linear causality. Indeed, it implies that an explanation in terms of causality is inappropriate – and perhaps even impossible. This is typically the case for systems that in part or in whole are intractable. The first known use of the term was by the English philosopher George Henry Lewes (1817–1878), who described emergent effects as not being additive and neither predictable from knowledge of their components nor decomposable into those components. In the contemporary vocabulary this means that the effects are non-linear and that the underlying system is partly intractable.

Resultant Outcomes

The way we typically explain how something has happened is by tracing back from effect to cause, until we reach the root cause – or run out of time and money. This can be illustrated by a representation such as the fishbone diagram shown in Figure 7.1.

Whatever happens is obviously real in the sense that there is an effect or an outcome, an observable change of something – sometimes described as a transition or a state change. It may be that the battery in a laptop is flat, that a report has been delivered, that a pipe is leaking, or that a problem has been solved. The outcome may also be a sponge forgotten in the abdomen of a patient, a capsised vessel, or a financial meltdown. Classic safety thinking goes on to assume that the causes are as real as the effects – which makes sense according to the logic that the causes in turn are the effects of other causes one step removed, and so on.

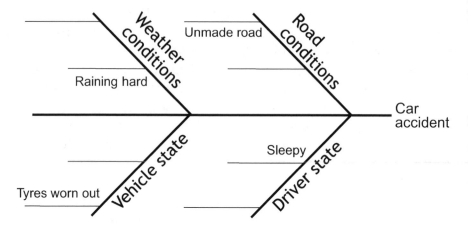

Figure 7.1   Fishbone diagram (resultant outcomes)

The purpose of accident and incident investigation is therefore to trace the developments backwards from the observable outcome to the efficient cause. Similarly, risk assessment projects the developments forwards from the efficient cause to the possible outcomes (cf., Figures 4.1 and 4.2).

A simple example is the investigation of a traffic accident, for instance that a car has veered off the road and hit a tree. An investigation may determine that the accident was due to a combination of the following: that the road was not well maintained, that the driver was sleepy and tired, that the tyres were worn out and that it was raining hard. In addition to assuming that these conditions existed when the accident happened, it may be possible to confirm or prove the assumptions empirically, because some of the conditions may also exist afterwards. The (wrecked) car can be inspected and the condition of the tyres established. The same goes for the road surface. We can find out how the weather was by consulting the meteorological office. And it may also be possible to determine the likelihood that the driver was sleepy by looking at what he or she had been doing for the preceding 12–24 hours. In other words, both the effects and the causes are 'real' in the sense that they can be independently verified and the outcome can therefore rightly be said to be a result of these causes. The reality of the causes also makes it possible to trace them even further back (following the same line of reasoning) and possibly do something about them, either eliminating them, isolating them, protecting against them (or rather their effects), etc.

Emergent Outcomes

In the case of emergent outcomes the causes are, however, elusive rather than real. The (final) outcomes are, of course, permanent or leave permanent traces, in the same way as for resultant outcomes. Otherwise it would be impossible to know that something has happened. The outcomes must furthermore be recognisable not only when they happen but also for some time afterwards – which often can be considerable. But the same need not be the case for whatever brought them about. The outcomes may be due to transient phenomena, combinations of conditions, or conditions that only existed at a particular point in time and space. These combinations and conditions may, in turn, be explained by other transient phenomena, etc. The combinations and conditions that are used to account for the observed outcomes

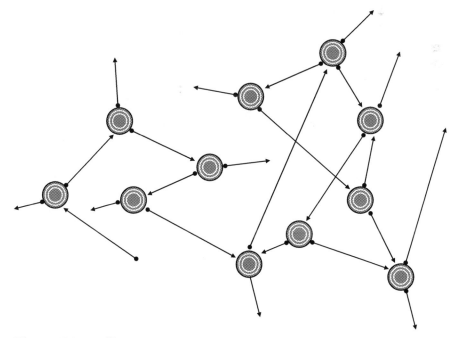

Figure 7.2 **Emergent outcomes**

are therefore constructed or inferred rather than 'found'. In the case of emergent outcomes, the 'causes' represent patterns that existed at one point in time but which did not leave any kind of permanent trace. The outcomes can therefore not be traced back to specific components or functions (Figure 7.2).

Emergent outcomes can be understood as arising from unexpected – and unintended – combinations of performance variability where the governing principle is resonance rather than causality. In relation to the ontology of Safety–II, this means that all the performance adjustments may be within an acceptable level or magnitude (which in practice means that they are too small to be noticeable), even though the outcome may be so large that is noticeable. This lack of proportionality between the precedents and the consequents is one reason why emergent outcomes are described as non-linear. It is also the reason why unwanted outcomes cannot be ensured by constraining variability, but only by controlling variability (monitoring and damping).

The principle of emergence means that while we may plausibly argue that a condition existed at some point in time, we can never be absolutely sure. The variability of everyday performance is, however, 'permanent' in the sense that there will always be some variability (cf., the discussion of the ontology of Safety–II). But since this performance variability – the everyday performance adjustments – does not constitute a failure or malfunction in the sense of the bimodality principle, it cannot be used as a selective cause to explain undesired outcomes. It is, on the other hand, systematic rather than random. This makes it largely predictable, which is why it can be used as the basis for safety analyses. Although we may not be able to do something about performance adjustments in the usual manner, we may be able to control the conditions that make them seem necessary, as long as these conditions are sufficiently regular and recurrent. We can, of course, also try to protect against these conditions by devising various forms of prevention and protection.

Resonance

Since emergence cannot be explained in terms of causality, and since we do need somehow to explain it – and not be satisfied with calling it 'emergence' – some practical principle is needed. Fortunately, a practical principle is also at hand, namely the idea of functional resonance. This has been presented and explained in another book, and interested readers are therefore referred to that. The Functional Resonance Analysis Method (FRAM) maps the couplings or dependencies among the functions of a system as they develop in a specific situation.

In physical systems, classical (or mechanical) resonance has been known and used for several thousand years. Resonance refers to the phenomenon that a system can oscillate with larger amplitude at some frequencies than at others. These are known as the system's resonant (or resonance) frequencies. At these frequencies even small external forces that are applied repeatedly can produce large amplitude oscillations, which may seriously damage or even destroy the system. In addition to classical resonance, stochastic resonance describes how random noise sometimes can push a subliminal signal over the detection threshold. (Stochastic resonance is of more recent origin and was first described in the beginning of the 1980s.) The outcome of stochastic resonance is non-linear, which simply means that the output is not directly proportional to the input. The outcome can also occur – or emerge – instantaneously, unlike classical resonance, which must be built over time.

Functional resonance differs from stochastic resonance because the emergent effect is based on approximate adjustments rather than random noise. The variability of performance in a socio-technical system represents the approximate adjustments of people, individually and collectively, and of organisations that together constitute everyday functioning. Because these approximate adjustments are purposive and rely on a small number of recognisable short cuts or heuristics, there is a surprisingly strong regularity in how people behave and in the way they respond to unexpected situations – including those that arise from how other people behave.

Performance variability is, indeed, not merely reactive but also – and perhaps more importantly – proactive. People not only respond to what others do but also to what they expect that others will do. The approximate adjustments themselves are thus made both in response to and in anticipation of what others may do, individually or collectively. What each person does obviously becomes part of the situation that others adjust to – again reactively and proactively. This gives rise to a dependence through which the functions in a system become coupled, and the performance variability of the functions therefore also becomes coupled. (People, of course, also respond to and in anticipation of how processes develop and what the technology may do; but in these cases there is no mutual adjustment, i.e., a process cannot 'expect' what the operator will do.)

Emergence can be found in all types of events, not only in the serious ones. The reason that it is more easily noted in serious events is that they are too complicated for linear explanations to be possible. This leaves emergence as the only alternative principle of explanation that is possible, at the moment at least. Emergence is nevertheless also present in many events that are less serious, but is usually missed – or avoided – because we only really put an effort into analysing the serious ones. Emergence thus forces its way to the front, so to speak, when it is impossible to find an acceptable single (or root) cause.

Once the use of emergence as an explanation has been accepted, a common reaction is for people to ask how emergent outcomes can be observed. But such a question represents a misunderstanding of the issue. Outcomes can certainly be observed, but whether they are emergent or not is a question of what types of explanations we can find and/or are happy with. Emergence is not something that can be seen, it belongs to the aetiology of Safety–II but not to the phenomenology.

The Phenomenology of Safety–II

Just as Safety–I was defined as a condition where as little as possible went wrong, Safety–II is defined as a condition where as much as possible goes right, indeed preferably as a condition where everything goes right. In analogy with resilience, Safety–II can also be defined as the ability to succeed under expected and unexpected conditions alike, so that the number of intended and acceptable outcomes (in other words, everyday activities) is as high as possible.

This definition of Safety–II leads to two questions. The first question is how – or why – things go right, or how we can understand why things go right. This question has already been answered by the ontology of Safety–II and, indeed, by the argument throughout this book that performance adjustments and performance variability is the basis for everyday successful performance. It has also been answered in several writings, such as the presentation of the ETTO principle.

The second question is how we can see what goes right. The phenomenology of Safety–II is all the possible outcomes that can be found in everyday work, the good as well as the bad and the ugly. The second question has already been addressed, first in the discussion of habituation in Chapter 3 and later in the discussion related to Figure 3.1. Here it was pointed out that it can be difficult to perceive things that go right because they happen all the time (habituation) and because we do not have a readily available vocabulary or set of categories by which to describe them.

The difficulty of perceiving that which goes right can be illustrated by a conversation between Sherlock Holmes and Inspector Gregory from Scotland Yard. Among the fictional detectives, Sherlock Holmes was famous for noticing things that others either did not see, or saw but found insignificant. The conversation can be found in the short story, 'Silver Blaze', which focuses on the disappearance of a famous racehorse on the eve of an important race and on the apparent murder of its trainer. During the investigation, the following conversation takes place:

> Gregory (a Scotland Yard detective): 'Is there any other point to which you would wish to draw my attention?'.
> Sherlock Holmes: 'To the curious incident of the dog in the night-time'.
> Gregory: 'The dog did nothing in the night-time'.
> Sherlock Holmes: 'That was the curious incident'.

The point here is that you can only notice – and know – that something is wrong if you know what should have happened in the everyday case. The understanding of how something is normally done (everyday work) is a necessary prerequisite for understanding whether something is (potentially) wrong. In this example Sherlock Holmes realised that a dog would normally bark when encountering a strange person. Since the dog had not barked (the 'curious incident'), the person who had entered the stable (and abducted Silver Blaze) had, therefore, not been a stranger.

Karl Weick famously proposed that reliability was a dynamic non-event (see Chapter 1), partly to explain why we do not notice (the effects of) reliability. For safety management, however, the non-event is far more important than the event. It is more important – or should be more important – that things go right than that things do not go wrong. The two states are not synonymous, the reason being that they are the result of quite different processes. Things go right because we try to make them go right, because we understand how they work and try to ensure that they have the best possible conditions to continue to do so. Things do not go wrong because we prevent them from going wrong, which means that we focus on the putative causes. In the former case, the starting point is a focus on successes – the Safety–II view, while in the latter it is a focus on failures – the Safety–I view.

In Safety–II the absence of failures (of things that go wrong) is a result of active engagement. This is not safety as a non-event, because a non-event can neither be observed nor measured. Safety–II is marked by a presence of successes (of things that go right), and the more there are, the safer the system is. In other words, safety is something that happens, rather than something that does not happen. Because it is something that happens, it can be observed, measured, and managed – in contrast to non-events. In order to ensure that a system is safe, we need, therefore, to understand how it succeeds rather than how it fails.

Safety–II: Ensuring That Things Go Right

As technical and socio-technical systems have continued to develop, not least due to the allure of ever more powerful information technology, systems and work environments have gradually become more intractable (Chapter 6). Since the models and methods of Safety–I assume that systems are tractable in the sense that they are well understood and well behaved, Safety–I models and methods are less and less able to deliver the required and coveted 'state of safety'. Because this inability cannot be overcome by 'stretching' the tools of Safety–I even further, it makes sense to change the definition of safety and to focus on what goes right rather than on what goes wrong (as suggested earlier in this book by Figure 3.1). Doing so will change the definition of safety from 'avoiding that something goes wrong' to 'ensuring that everything goes right' or, in a paraphrase of the definition of resilience, to the ability to succeed under varying conditions, so that the number of intended and acceptable outcomes (in other words, everyday activities) is as high as possible. The consequence of this definition is that the basis for safety and safety management now becomes an understanding of why things go right, which means an understanding of everyday activities.

Safety–II explicitly assumes that systems work because people are able to adjust what they do to match the conditions of work. People learn to identify and overcome design flaws and functional glitches because they can recognise the actual demands and adjust their performance accordingly, and because they interpret and apply procedures to match the conditions. People can also detect and correct when something goes wrong or when it is about to go wrong, so they can intervene before the situation becomes seriously worsened. The result of that is performance variability, not in the negative sense where variability is seen as a deviation from some norm or standard, but in the positive sense that variability represents the adjustments that are the basis for safety and productivity (Figure 7.3).

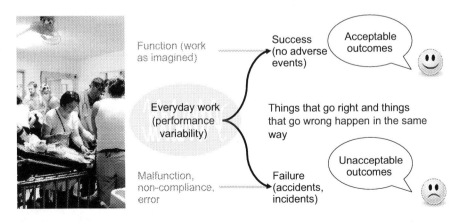

Figure 7.3   The Safety–II view of failures and successes

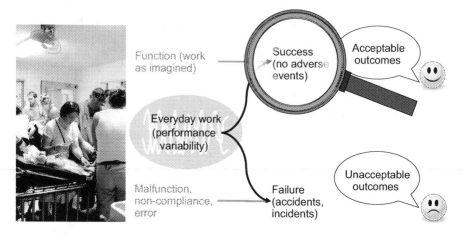

Figure 7.4   Understanding what goes wrong by understanding what goes right

Performance variability or performance adjustments are a sine qua non for the functioning of socio-technical systems, unless they are extremely simple. Therefore unacceptable outcomes or failures cannot be prevented by eliminating or constraining performance variability, since that would also affect the desired acceptable outcomes (cf., the hypothesis of different causes in Chapter 3). Instead, efforts are needed to support the necessary improvisations and performance adjustments by understanding why and how they are made, by clearly representing the resources and constraints of a situation and by making it easier to anticipate the consequences of actions. The somewhat paradoxical consequence for investigations of adverse events in Safety–II is, therefore, that first, one should study the situation when things went right (Figure 7.4); this is discussed further in Chapter 8 as 'breadth before depth'. Performance variability should be managed and dampened if it looks like it is going in the wrong direction, or amplified if it looks like it is going in the right direction. In order to do this it is necessary first to recognise performance variability, second to monitor it and third to control it. That is the remit of safety management according to Safety–II.

Proactive Safety Management

Safety–II management and resilience engineering both assume that everything basically happens in the same way, regardless of the outcome. This means that it is unnecessary to have one set of causes and 'mechanisms' for things that go wrong (accident and incidents), and another for things that go right (everyday work). The purpose of safety management is to ensure the latter, but achieving this will also reduce the former. Although Safety–I and Safety–II both lead to a reduction in unwanted outcomes, they use fundamentally different approaches with important consequences for how the process is managed and measured and for productivity and quality.

From a Safety–II perspective, safety management cannot achieve its stated purposes by responding alone, since that will only correct what has happened. Safety management must instead be proactive, so that adjustments are made *before* something happens, which will therefore affect how it happens, or even prevent something from happening. A main advantage is that early responses, on the whole, require less effort because the consequences of the event will have had less time to develop and spread. And early responses obviously save valuable time.

For proactive safety management to work, it is necessary to foresee what can happen with acceptable certainty and to have the appropriate means (people and resources) to do something about it. That in turn requires an understanding of how the system works, of how its environment develops and changes, and of how functions may depend on and affect each other. This understanding can be developed by looking for patterns and relationships across events rather than for causes of individual events. To see and find those patterns, it is necessary to take time to understand what happens and not spend all resources on firefighting.

A trivial example is to 'batten down the hatches' when bad weather is approaching. While this expression has its origin in the navy, many people living on land (in a tornado zone, for instance) or on an oil rig have also learned the value of preparing for a storm. In the financial world, proactive safety management is de rigueur; and a financial institution that can only react will soon be out of business. In a different domain, the precautions following the World Health Organization's warning in 2009 of a possible H1N1 flu pandemic offer another example of proactive safety management. After the warning was issued, European and other governments began to stockpile considerable amounts of drugs and vaccines to ensure that the necessary resources were in place. Although the warning later turned out to have been a false alarm, it illustrates the essential features of proactive safety management.

It is obviously a problem for proactive safety management that the future is uncertain and that an expected situation may fail to happen. In that case, preparations will have been made in vain and time and resources may have been wasted. It is also a problem that predictions may be imprecise or incorrect, so that the wrong preparations are made. Proactive safety management thus requires taking a risk, not least an economic one. But the alternative of not being ready when something serious happens will indubitably be even more expensive in both the short and the long run.

On Monolithic Explanations

Safety–I thrives on monolithic explanations, such as situation awareness, safety culture, 'human error', and the like. These causes explain what has gone wrong and therefore typically refer to the absence of something: lack of situation awareness, lack of safety culture, etc. And although 'human error' is noted by the presence of error, in the forms of mistakes, violations, etc., depending on the preferred nomenclature, this is actually the absence of something, namely the absence of error-free performance. These explanations are similar because they all rely on the existence of causes that furthermore should be simple and single. In contrast to that, Safety–II tries to explain and understand how things work, how everyday work is done. Safety–II therefore looks less for causes ('X went right because of Y'), and more for a way to describe how work is done.

It may be objected that performance variability, or for that matter the various forms of trade-offs, also represent single explanations. But that is only the case from a very superficial view. Performance variability is a characteristic of performance, whether it goes right or goes wrong. It therefore does not explain performance – or it explains all kinds of performance, which makes it useless as a cause. By acknowledging that performance is variable, that it is based on multiple approximate adjustments, we can begin to look at how these happen and use that as a way to understand Work-as-Done. There are multiple adjustments for multiple reasons, rather than just one type of adjustment for one reason.

Are There No Simple 'Errors'?

One of the comforts of the *causality credo*, combined with the myths of Safety–I, is that it becomes easy to find simple – and sometimes oversimplified – explanations for things that have gone wrong. This convenience seems to be lost with the emphasis on performance adjustments and performance variability as a basis for Safety–II.

Although Safety–II replaces the bimodality principle by the principle of approximate adjustments, it does not mean that there are no longer simple 'errors', i.e., that people never do something that is incorrect for simple reasons. Indeed, the majority of uncomplicated cases may still be understood and explained in this way without missing anything significant.

Consider, for instance, a new worker or a novice. A novice is by definition someone who has little or no experience with how a certain task should be done or how a piece of equipment should be used. There will therefore be many situations that are unknown or unrecognised, where the novice struggles to find out what to do. It is therefore less a question of adjusting performance to the conditions than a question of knowing how to do something elementary. In such cases people may resort to trial and error – to opportunistic control – and therefore happen to do things that clearly are incorrect and so can be considered simple 'errors'. The dilemma is that a novice can either try to be thorough and find out what the proper action is by searching for information, asking others, etc., or try to be efficient and make a decision based on incomplete understanding or some kind of rule of thumb.

There are also simple 'errors' in the sense of imprecision and performance variability induced by fatigue. While imprecision in some sense may be understood as an ETTO, there are clearly situations where we do something that is wrong, for instance due to a speed–accuracy trade-off, because the environment is shaking or unstable (as in a car or on a ship), because the interface is badly designed, illegible, or ambiguous, and for a host of other reasons. This can be seen as exogenic and endogenic performance variability, and is in some sense irreducible. For the exogenic variability, particularly for single actions, it behoves the designer to consider this and make sure that incorrect actions can be caught, as in the famous confirmation dialogue box on computer screens.

Comments on Chapter 7

The 'law of unintended consequences' is a lucid analysis of why formally organised action sometimes fails to achieve its objectives. The original paper is in Merton, R.K. (1938), The unanticipated consequences of purposive social action, *American Sociological Review*, 1(6), 894–904). Despite having been written 75 years ago, the analysis is unfortunately still valid.

There has always been a strong belief in the potential of technology to overcome human limitations with regard to, for example, force, speed, precision, perception and cognition. This belief has made automation a preferred solution to many human factors problems. In a seminal analysis of the use of automation to replace humans, Lisanne Bainbridge pointed out some of the fundamental limitations in a delightful paper published in 1983 entitled 'Ironies of automation', *Automatica*, 19, 775–79.

The adjustments that people need to make in order to be able to do their work have frequently been discussed in the ergonomics literature. A particularly interesting paper is Cook, R.I. and Woods, D.D. (1996), Adapting to new technology in the operating room, *Human Factors*, 38(4), 593–613, where an important distinction is made between system tailoring and task tailoring. System tailoring describes how users try to change the system setup and interface so that work becomes possible. When system tailoring is limited or impossible, users resort to task tailoring, which involves changing or adjusting the tasks so that work becomes possible.

The ontology of Safety–II refers to socio-technical systems and especially to human activities, as individuals, as social groups and as organisations. And humans are clearly not bimodal. As far as the other 'half' of socio-technical systems is concerned – the technology – this can be considered as bimodal. But this does not change the fact that the socio-technical system as a whole is variable rather than bimodal.

The term 'emergence' was used by George Henry Lewes in the debates about Darwin's theory of evolution. At that time machinery was clearly understandable, and emergence was not a problem in relation to work situations.

The role of functional resonance in how a work situation may develop has been described in Hollnagel, E. (2012), *The Functional Resonance Analysis Method for Modelling Complex Socio-technical Systems*, Farnham: Ashgate. The method FRAM has specifically been developed to understand the importance of performance variability in work. The book includes a discussion of the three types of resonance: classical resonance, known since the Athenian culture; stochastic resonance, first described in the beginning of the 1980s; and finally functional resonance, which is part of the aetiology of Safety–II.

The Sherlock Holmes story, 'Silver Blaze', is in *The Memoirs of Sherlock Holmes* by Arthur Conan Doyle. The text can be downloaded from the Gutenberg project (http://www.gutenberg.org).

Chapter 8
The Way Ahead

Consequences of a Safety–II Perspective

Adopting a Safety–II perspective does not mean that everything must be done differently or that currently used methods and techniques must be replaced wholesale. The practical consequence is rather a recommendation to look at what is being done in a different way. It is still necessary to investigate things that go wrong, and it is still necessary to consider possible risks. But even a root cause analysis can be done with another mindset, and even a fault tree can be used to think about variability rather than probability.

While day-to-day activities at the sharp end never are reactive only, the pressure in most work situations is to be efficient rather than thorough. This pressure exists at all levels of an organisation and inevitably reduces the possibilities of being proactive because being proactive requires that some efforts are spent up front to think about what could possibly happen, to prepare suitable responses, to allocate resources, and make contingency plans.

Even when proactive safety management is in place – at least as a policy or an organisational attitude – it can be difficult to be proactive for the myriad small-scale events that constitute everyday work situations. Here, things may develop rapidly and unexpectedly with few leading indicators. The overriding concern will be to keep things running smoothly with resources that often are stretched to the limit. There may therefore both be fewer resources to allocate and less time to deploy them. The pace of work leaves little, if any, opportunity to reflect on what is happening and, therefore, little possibility of being tactical – to say nothing of being strategic. Instead, work pressures and external demands require quick-and-dirty solutions that may force the system into an opportunistic mode where it is more important that responses are quick than that they are accurate. To get out of this – to switch from an opportunistic to a tactical control mode and thereby begin to become proactive – requires a deliberate effort. While this might not seem to be affordable or justifiable in the short term, it is unquestionably a wise investment in the long term.

It is somewhat easier to be proactive for large-scale events because they develop relatively slowly – even though they may often begin abruptly. (An example would be the eruption of a volcano that may lead to the closure of airspace, or the sudden outbreak of a pandemic disease.) The further development of such large-scale events is more likely to be regular rather than irregular, provided no other similar event occurs, and there are often reliable indicators for when a response is needed. The appropriate responses are furthermore known, so that preparations in principle can be made ahead of time.

While Safety–II represents an approach to safety that in many ways differs from Safety–I, it is important to emphasise that they represent two complementary views of safety rather than two incompatible or conflicting views. Many of the existing practices can therefore still be used, although possibly with a different emphasis. But the transition to a Safety–II view will also require some new practices, as described below. Juxtaposing Safety–I and Safety–II is therefore useful to draw attention to the consequences of having one or the other as the basis for safety management, cf. Table 8.1.

What people usually do in everyday work situations at the sharp end can be described as a judicious mixture of a Safety–I and a Safety–II approach. The simple reason is that effective performance requires both that people can avoid that things go wrong and that they can ensure that things go right. The specific balance depends on many things, such as the nature of the work, the experience of the people, the organisational climate, management and customer demands, production pressures, etc. Everybody knows that prevention is better than cure, but the conditions may not always allow prevention to play its proper role.

Table 8.1   A comparison of Safety–I and Safety–II

	Safety–I	Safety–II
Definition of safety	As few things as possible go wrong.	As many things as possible go right.
Safety management principle	Reactive, respond when something happens, or is categorised as an unacceptable risk.	Proactive, continuously trying to anticipate developments and events.
Explanations of accidents	Accidents are caused by failures and malfunctions. The purpose of an investigation is to identify causes and contributory factors.	Things basically happen in the same way, regardless of the outcome. The purpose of an investigation is to understand how things usually go right as a basis for explaining how things occasionally go wrong.
Attitude to the human factor	Humans are predominantly seen as a liability or a hazard.	Humans are seen as a resource necessary for system flexibility and resilience.
Role of performance variability	Harmful, should be prevented as far as possible.	Inevitable but also useful. Should be monitored and managed.

It is a different matter when it comes to the levels of management and regulatory activities where the Safety–I view dominates. One reason is that the primary objective of management and regulators historically has been to prevent workers, customers and the public from being exposed to unacceptable hazards. Management, regulators and lawmakers are, however, removed in time and space from the daily operations of the systems and services they are in charge of and therefore have limited opportunity to observe or experience how work actually is done. This makes it inevitable that the focus is on what has gone wrong or may go wrong and how it can be prevented. Another reason is that it is much simpler for everyone to count the few events that fail than the many that do not – in other words an efficiency–thoroughness trade-off. (It is also – wrongly – assumed to be easier to account for the former than for the latter.)

Since the socio-technical habitats on which our daily life depends continue to become more and more complicated, remaining with a Safety–I approach will become inadequate in the long run, if that has not already happened. Complementing current safety management practices with a Safety–II perspective should therefore not be a difficult choice to make. The recommended solution is not a wholesale replacement of Safety–I by Safety–II, but rather a combination of the two ways of thinking. It is still the case that the majority of adverse outcomes have relatively simple explanations – or at least that they can be treated as having relatively simple explanations without serious consequences – and therefore that they can be dealt with in ways we have become accustomed to and know well. But it is also the case that there is a growing number of situations where this approach will not work. Since many practices of Safety–I already have been stretched to or beyond their breaking point, it is necessary to find an alternative – which essentially means adding a Safety–II perspective and moving towards a resilience engineering practice, cf. Figure 8.1. Safety–II and resilience engineering together represent a different way of looking at safety, corresponding to a different definition of what safety is, hence also a different way of applying known methods and techniques. In addition to that, a Safety–II perspective will also require methods and techniques on its own to be able to look at things that go right, to be able to analyse how things work, and to be able to *manage* performance variability rather than just *constraining* it. The most important of these methods and techniques are described in the following sections.

Figure 8.1   Relationship between Safety–I and Safety–II

Looking For What Goes Right

For Safety–II it is important to look at what goes right as well as what goes wrong, and to learn from what succeeds as well as from what fails. Indeed, it is essential not to wait for something bad to happen, but to try to understand what actually takes place in situations where nothing out of the ordinary seems to take place. Safety–I assumes that things go well because people simply follow the procedures and work as imagined. Safety–II assumes that things go well because people always make what they consider sensible adjustments to cope with current and future situational demands. Finding out what these adjustments are and trying to learn from them can be more important than finding the causes of infrequent adverse outcomes!

When something goes wrong, such as a runway incursion, driving a train too fast in a curve, or overlooking a test response, the activity involved is unlikely to be unique. It is more likely to be something that has gone well many times before and that will go well many times again in the future. From a Safety–II view adverse outcomes do not happen because of some kind of error or malfunction, but because of unexpected combinations of everyday performance variability. It is therefore necessary to understand how such everyday activities go well – how they succeed – in order to understand how they fail.

A simple but powerful example of this is a case described in H.W. Heinrich's 1931 book, *Industrial Accident Prevention*. The adverse outcome or injury was that a mill employee fractured his kneecap because he slipped and fell on a wet floor. It had, however, for more than six years been the practice to wet down too great an area of floor space at one time and to delay unnecessarily the process of wiping up. Slipping on the part of one or more employees was therefore a daily occurrence, although usually without serious consequences. (The ratio of no-injury slips to the injury was 1,800:1.) Since Heinrich does not relate what happened after the accident, it is anybody's guess whether the established practice of floor cleaning was changed.

The difference between a Safety–I and a Safety–II view can be illustrated as shown in Figure 8.2. Safety–I focuses on events at the tails of the normal distribution, and especially on the infrequent

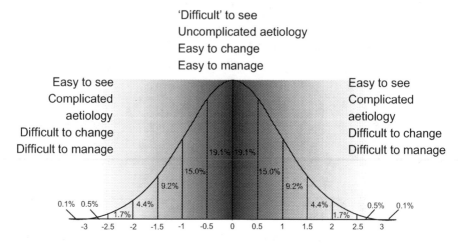

Figure 8.2   Relationship between event probability and ease of perception

events with negative outcomes at the left tail that represent accidents. (Curiously, there seems to be little interest for the infrequent events at the right tail that go exceptionally well.) Such events are easy to see because the outcomes have noticeable effects that differ from the usual. They are, however, difficult to explain and understand – the attractiveness of root causes and linear models notwithstanding. Explanations that are based on linear causality are, however, dangerously oversimplified. Because the events are rare and because they are difficult to explain and understand, they are also difficult to change and manage. For proof of that one need look no further than to the ineffectiveness of commonly used recommendations such as 'follow procedures', 'be more careful', 'improve training', etc.

Safety–II focuses on the frequent events in the middle of the distribution – as well as events at the tails. These frequent events are 'difficult' to see, but mainly because we habitually ignore them. Given that resources always are limited, the 'logic' seems to be that if something works, then there is no reason to spend more time on it. But the fact of the matter is that things usually do not work in the way that we assume, and that Work-As-Done is significantly different from Work-As-Imagined. The events in the middle of the distribution can be understood and explained in terms of the mutual performance adjustments that constitute the basis for everyday work. Because these events are frequent, because they are small scale, and because we can understand why and how they happen, they are easy to monitor and manage. Interventions can be precise and limited in scope (because the subject matter is uncomplicated), and it is therefore also easier – although not necessarily straightforward – to anticipate what both the main and the side effects may be.

There is of course an evolutionary benefit in not paying attention (or at least not too much attention) to the usual as long as it does not harm us and as long as the environment is stable. But in our current socio-technical habitats, the environment is no longer stable. Even without subscribing to the notion of Complex Adaptive Systems, work environments and, therefore, also work itself, have become increasingly difficult to understand and so less predictable. In consequence of that, solutions and routines that serve us well today may not serve us well tomorrow. It is therefore important not just to rely on routines but to pay attention to what actually takes place. That is the kind of thoroughness that will enable us to be efficient when the time comes to make changes, and to make them rapidly.

Breadth Before Depth

Looking for what goes right requires that the analysis prioritises breadth before depth. Accident analyses and event investigations in Safety–I treat each event as unique, and tries to find the specific causes for what happened. This leads to a search that is characterised by depth-first or 'depth before breadth', meaning that each possible path leading to the outcome is explored as far back as possible, before alternatives are considered. One obvious drawback of this is that the search may be stopped when an acceptable cause has been found. A depth-first analysis will therefore get stuck with 'first stories' and fail to consider 'second stories'. Even if more paths are explored, the event is still treated as unique rather than as a possible variation of a common way of doing things. Another drawback is that once an event has been analysed and a set of recommendations made, the problem is considered solved and the case closed. In this way learning is discouraged, and it is rare that any efforts are made to consider a set of events together. The past is only reviewed when a new accident makes it unavoidable.

When an event is analysed from a Safety–II perspective, it is taken for granted that the event is not unique but rather something that has happened before and that will happen again. The analysis therefore begins by considering the event as representative of everyday work, and tries to understand what characterises that – the typical conditions that may exist and the adjustments that people learn to make. This leads to a breadth-first or 'breadth before depth' approach, where the many ways in which the event can take place are explored before going into depth with any of them. The first cause, or 'first story', found is therefore less likely to be considered as final, because it is obvious that this is just one way out of many in which things could happen.

Finding What Goes Right

When faced with the prospect of finding what goes right, the task is daunting, You may, of course, simply begin to look at what others do every day – or, even better, pay attention to what you do yourself. It is actually not so difficult to do so and the basic vocabulary has been presented in the preceding section. (See also 'Types of Performance Adjustments' on page 156.)

Most people have little or no practice in just looking at what happens, and it may therefore be useful to make a deliberate effort, at least initially. Work descriptions often start from ideas about how an activity ought to be carried out – for instance, as they can be found in design documents, instructions, procedures, training materials, etc. Looking at what happens is, however, about Work-As-Done rather than about Work-As-Imagined and must therefore refer to work as it is usually is done in an everyday setting.

The best source of information for this is not surprisingly the people who actually do the work, either at the workplace that is directly associated with the analysis, or at a workplace that is highly similar. The primary source for getting this information is interviews; a secondary source may be field observations, or even an exchange of people between departments or units, since this will provide a set of fresh eyes. The discussion here will nevertheless limit itself to systematic data collection by interviews.

Before conducting an interview it is important carefully to think through the situation and to consider how the information is going to be used. It is, as always, important to prepare well before going into 'the field', for instance by consulting available sources of information, such as rules and regulations, statistics for various types of events, known 'worst cases' or 'worst scenarios', stability of the workplace (rate of change of staff, equipment, procedures, organisation), and the commonly known history of major events or changes (preferably not limited to accidents) that have happened in the near past. This background information is the basis for defining the set of questions that should be asked during the interviews.

It is equally important to know as much as possible about the workplace itself, i.e., the actual physical and environmental conditions (or context) where work takes place. This information can be found by looking at architectural drawings (layout of the workplace), photos and videos, and other available types of information. The data collection/interview should also – if at all possible – take place at the actual place where the activity is carried out. A 'guided tour' of the premises is an additional source of valuable information that is not easily conveyed in any other form. Walking around to get a sense of what it is like to work in a particular setting is very useful both for asking questions and for interpreting answers. A good interviewer will bring a pair of 'fresh' eyes to the setting and may notice things that the people who work there no longer see.

The goal of an interview is to find out how people do their work. This can be prompted by some simple questions such as:

- When do you typically start the (specified) activity? What is it that 'activates' it?
- Do you ever adjust or customise the activity to the situation? How? How do you determine which way to proceed?
- What do you do if something unexpected happens? For example, an interruption, a new urgent task, an unexpected change of conditions, a resource that is missing, something that goes wrong, etc.

- How stable are the working conditions? Is your work usually routine or does it require a lot of improvisation?
- How predictable is the work situation and the working conditions? What can happen unexpectedly and how do you prepare for it?
- Is there something that you often have to tolerate or get used to during everyday work?
- What preconditions for your work are usually fulfilled? Are these preconditions shared with all who take part in the work?
- Are there any factors which everyone takes for granted during work?
- How do you prepare yourself for work (e.g., reading documents, talking to colleagues, refreshing instructions, etc.)?
- What data do you need? What kind of equipment, apparatus, or service features do you need? Can you safely assume that they will be available when needed?
- What do you do in case of time pressure?
- What do you do if information is missing, or if you cannot get hold of certain people?
- What skills do you need?
- What is the best way to perform your work? Is there an optimal way to do it?
- How often do you change the way you work (rarely, often)?

It is also essential to prepare the people who are interviewed. First of all, they must of course agree to be interviewed. Once this agreement has been obtained, it is important that they are informed about what the purpose and nature of the data gathering is. The general experience is that people are more than willing to tell about how they do their work, and how they manage tricky situations. It may be a good idea to interview two people at the same time, since they then often realise that one person may do things quite differently from another.

In addition to interviews – and field observations – some general techniques from organisational development may also be used. They all share the position that the questions we ask tend to focus the attention in a particular direction, a version of 'what you look for is what you find'. Looking for how work is done, rather than for how something went wrong, will produce different types of information and potentially even change people's mindset – to say nothing about the organisational culture. The search should be for how problems are solved rather than for which problems there are.

The best known of these methods is probably *Appreciate Inquiry*. This method focuses on understanding what an organisation does well rather than on eliminating what it does badly – in line with the difference between Safety–II and Safety–I. Even though it is not specifically aimed at safety, it can be used to develop a more constructive culture, which presumably is correlated with safety performance. The possible downside of *Appreciate Inquiry* is that it requires extensive time (often workshops stretching over several days), and that it only includes a limited number of people.

Another method is called *Co-operative Inquiry*. Its concrete guiding principle is that people are self-determining, meaning that people's intentions and purposes are the causes of their behaviour. It is important to facilitate that by exploring and acknowledging the thinking and decision-making that allow people to work constructively together. Co-operative Inquiry and Safety–II have in common that they accept performance variability as a useful contribution to work. For both Co-operative Inquiry and Appreciate Inquiry it is, however, more important to improve the work culture or work climate, than address the product or the outcome, i.e., the approaches are more therapeutic than operational.

A third, and for now final, approach goes by the name of *exnovation*. The focus of exnovation is on what practitioners do in their day-to-day work, not framed in statistics but described as far as possible as it happens or as it is experienced. The focus is on how practitioners together engage in developing and maintaining a practice, on how they learn about each other, about the issues, and about the futures they could seek. A prerequisite for such learning is 'passive competence', defined as the ability to put typical conducts and answers on hold, to defer routine responses, to query existing practice and knowledge and to legitimise new kinds of answers and solutions. There are thus strong similarities to the principle of mindfulness described further down. This 'passive competence' and openness are critical for engaging with the complicated details of what goes on, without prejudging it or becoming trapped by routines and moves that are inappropriate for the situation. This underscores that exnovation is not in the first instance about identifying knowledge, rules or truths, but about partaking in a deliberative process, in a forum of engagement.

Types of Performance Adjustments

When something is done as part of daily work, it is a safe bet that it has been tried before, as is argued above. People are good at finding ways of doing their work that meet the requirements – their own level of ambition, the social expectations, and the organisation's demands – and overcome the permanent and temporary problems that exist in every work situation. Because these performance adjustments work, people quickly come to rely on them – precisely because they work. Indeed, they may be tacitly reinforced in this when things go right but blamed when things go wrong. Yet blaming people for doing what they usually do is short-sighted and counterproductive. It is more conducive to both safety and productivity to try to find the performance adjustments people usually make, as well as the reasons for them. This has two obvious advantages. The first is that it is far easier and less incriminating to study how things go right than to study things that have gone wrong. The second is that trying to ensure that things go right by monitoring and managing performance adjustments is a better investment of limited resources than trying to prevent failures.

There is, however, a practical problem, namely that a terminology is not readily available. When we want to describe various types of individual and organisational failures, errors and malfunctions, a rich vocabulary is at our disposal. Starting from simple categories such as errors of omission and commission, we not only have multiple theories (violations, non-compliance, loss of situation awareness, cognitive errors, error-forcing conditions, etc.) but also a rich flora of terms within each theory. But when we want to describe what people actually do, there is little to start from.

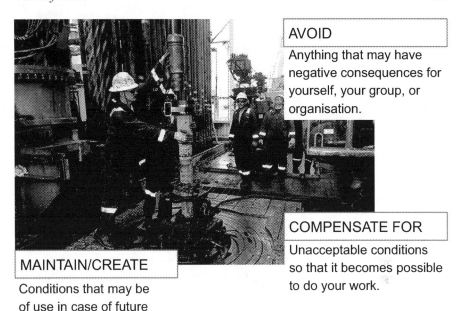

AVOID
Anything that may have negative consequences for yourself, your group, or organisation.

COMPENSATE FOR
Unacceptable conditions so that it becomes possible to do your work.

MAINTAIN/CREATE
Conditions that may be of use in case of future problems.

Figure 8.3   Types of performance adjustments

In the spirit of Safety–II, one thing we can look for is how people adjust their performance to the situation. The concrete adjustments will obviously have to be appropriate for the work environment and it is therefore impossible to provide a set of universally applicable categories. Instead, one can look at the reasons why people adjust their performance and from that propose some characteristic types of adjustments that can easily be observed. If we do this, it turns out that people in practice adjust their performance for three main motivations: to maintain or create conditions that are necessary for them to do their work or that may be of use in case of future problems, to compensate for something that is missing, and to avoid future problems, cf. Figure 8.3.

Maintaining or Creating Good Working Conditions

For every work situation certain conditions must be fulfilled. These can relate to the tools and equipment, to interface design, to the availability of materials and information, to time available and workload (or the absence of time pressures and overload), to protection against disturbances and interruptions, to access to colleagues and experts, to ambient working conditions (light, heat, humidity, noise, vibrations, etc.), to working positions and working hours, to the scheduling and synchronisation of different phases of work or different functions, etc. For every work situation efforts are also made in advance to ensure that these conditions are met, nominally as well as actually. Indeed, the more stringent the demands to work output are, the more effort is put into ensuring that these conditions are in place. Yet it is also true that it is impossible to ensure that the necessary conditions always are fulfilled. One type of adjustment that people make are therefore to establish or bring about these conditions, or to maintain them. People do so because they are fully aware that such problems might adversely affect their work, and that they therefore need to do something about them.

Possibly the best example of that is the management of time. We always need sufficient time to do our work and we therefore try to anticipate and prevent interruptions and disruptions. This can be as simple as putting a 'do not disturb' sign on the door (or on the Skype profile), or a bit more complicated such as trying to minimise disturbances by imposing restrictions on what others can do. We furthermore often try to preserve or create a little time by working a bit faster (which means making other adjustments) or by not replying to every email or responding to every phone call immediately.

Compensating for Something that is Missing

Even with the best of intentions and with the best of planning, there will inevitably be situations where something is missing, where one or more conditions are unacceptable, so that it becomes difficult or impossible to carry out the work. Time is, of course, again an important concern, but there are others. A tool or a piece of equipment may be missing, materials may not be available or be available in the wrong form (dose, concentration) or volume (enough cement to plug a blow-out, enough water or retardants to extinguish a fire, or enough available beds to accept all incoming patients). Information may be missing (information underload) and it may be impossible to get additional information in time – or at all; equipment may break down; or people may not be available when needed. In such cases we must make the appropriate adjustments or fail the task. If time is too short we try to do things faster, to compensate for the lack of time. If tools or equipment are missing, we find some substitute – a trivial example being to use a screwdriver as a hammer. If a material is missing or in short supply, we try to find substitutes or to reduce the demands. If information is missing we try our best guesses, for instance through frequency gambling or similarity matching, and so on. Regardless of what is missing, we usually find a way of compensating for it.

Avoiding Future Problems

Finally, adjustments can be made to avoid anything that may have negative consequences for the person, the group, or the organisation. The execution of a procedure may be changed or rescheduled to avoid some consequences (such as interfering with other people at the workplace); barriers may be established to prevent an impending disturbance or disruption; a piece of equipment may be cannibalised to avoid a temporary cessation of activity; a report may be withheld to delay outside interference by regulators, lawmakers, or the press; an activity may be postponed to await more suitable conditions, etc. It makes sense to adjust how things are done if continuing unchanged would lead to trouble; indeed, it would – in hindsight – be reckless not to do so.

We all more or less constantly use all three types of adjustments, either alone or in combination. We apply them so fluently and effortlessly that we rarely notice that we – or others – do so. Because they are an integral part of Work-As-Done, we expect them to be made and often anticipate that they will be, though more often tacitly than explicitly. By using the simple categories described above, we can begin to notice the adjustments when they are made and thereby gain an understanding of the characteristic performance variability in a specific situation or activity.

The Basis for Learning: Frequency vs Severity

Everyone who is involved with efforts to improve safety just knows that learning should be based on accidents, on things that have gone wrong. Accidents are accepted as opportunities to learn, specifically to find out what should be done to make sure that the same – or a similar – accident will not happen again. Since accident investigations often are limited by time and resources, there is a tendency to look at the accidents that have serious consequences and leave the rest to be dealt with later – although that 'later' rarely comes. The same line of thinking goes for incidents and is perhaps even more pronounced here, since the number of incidents is usually very large. The unspoken assumption is that the potential for learning is proportional to the severity of the incident/accident. This is obviously a mistake – and is also in conflict with the principles of the pyramid model discussed in Chapter 4. While it is correct that more money can be saved by avoiding a major accident than by avoiding a minor one, it does not mean that the learning potential is greater as well. Indeed, it might easily be the case that the accumulated cost of frequent small-scale incidents is larger than the cost of a single infrequent accident. Small but frequent events are furthermore easier to understand and easier to manage (cf., above). It therefore makes better sense to base learning on those than on rare events with severe outcomes.

Another argument for using frequency rather than severity as a starting point for learning comes from general learning theory. Considered quite generally, whether it is related to safety or to something else, three conditions must be fulfilled in order for learning to take place:

1. There must be an opportunity to learn. Situations where something can be learned must be frequent enough for a learning practice to develop. They must also happen so frequently that what has been learned is not forgotten in the meantime. Since accidents are infrequent, they are not a good starting point for learning. Because they are infrequent, lessons may be forgotten between opportunities to learn. For the same reason it is also difficult to be prepared to learn –

when an accident happens, the main concern is to deal with it and to restore the system, rather than to learn from it.

2. In order for learning to take place, situations or events must be similar in some way. Learning situations must have enough in common to allow for generalisation. If events are very similar, then it is easy to learn, because we have an innate ability to generalise. But if events are very dissimilar – as serious accidents are – or if it is impossible for other reasons to generalise (to find something in common), then it is more difficult to learn. We can clearly not learn just from the outcomes, since similar or identical outcomes may have different causes. We need to go beyond that and learn from the causes. But the causes are inferred and constructed rather than observed. In the case of simple events, there may in practice be a one-to-one correspondence between cause and effect because there are few links or steps in between. But for a complicated event that is no longer the case. The more serious an accident is, then – usually – the more steps there are on the way from the outcome to the causes, and the more details in the explanations, hence more uncertainty. This means that we can only learn by postulating some kind of underlying causal mechanism, and what we then learn is that the 'mechanism' is good enough to provide an explanation (that is socially acceptable), but not necessarily what really happened. The question, however, is whether we should aim to reinforce our belief in a specific accident model or a theory, or whether we should try to understand what actually goes on and generalise from that?

3. There must also be an opportunity for feedback, both to demonstrate that something has been learned, and to verify that what has been learned is relevant and effective. A prerequisite for that is that events are both frequent and similar. If neither is the case, then it is difficult to be sure that the right thing has been learned. Learning is not just a random change in behaviour but a change that is supposed to make certain outcomes more likely and other outcomes less likely. It must therefore be possible to determine whether or not the learning (i.e., the desired change in behaviour) has occurred and whether it has the expected effects. If there has

been no change, then learning has not been effective. And if there has been a change in the opposite direction, then learning has certainly been wrong.

We can illustrate the consequences of these three criteria by mapping situations according to their frequency (how often they occur) and their similarity (how similar two events are). If we simply use the three major categories of accidents, incidents, and everyday performance, the mapping looks as shown in Figure 8.4. (Note, by the way, that this can be seen as including the traditional accident pyramid turned 135 degrees counterclockwise.) General learning theory confirms what we already knew, namely that the best basis for learning is found in events that happen often and that are similar. Learning should therefore be based on frequency rather than severity, which means that accidents are neither the best nor the only basis for learning.

In relation to learning, the difference between Safety–I and Safety–II may also be expressed as follows. The purpose of investigations in Safety–I is to find what went wrong, while the purpose of investigations in Safety–II is to find what did not go

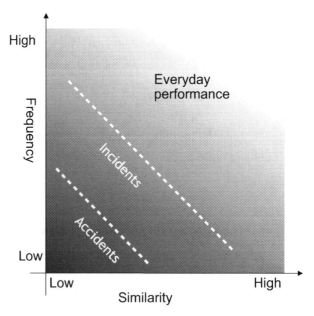

Figure 8.4   What does it take to learn?

right. The first is done by reconstructing the assumed failure sequence (or accident timeline), and then finding the component or subsystem that malfunctioned or failed. This naturally makes severity the most important criterion. The second is done by constructing an account of everyday, successful performance and then looking for how performance variability, alone or in combination, could lead to loss of control. This just as naturally makes frequency the most important criterion. Asking for what went *wrong* leads to a search for errors, failures and malfunctions. Asking for what *did not go right* creates a need to understand how work normally takes place and how things go right. Only when that has happened will it be possible to understand what did not go right in the specific instance.

To say that learning should be based on frequency rather than severity is, however, not an argument against looking at accidents and incidents. It is rather an argument against *only* learning from accidents and incidents. It is essential to learn from what happens every day – from performance variability and performance adjustments – because this is the reason why things sometimes go wrong, and because this is the most effective way to improve performance.

Remain Sensitive to the Possibility of Failure

The gradual identification and elimination of things that go wrong slowly creates an attitude that the situation is under control. To counter that it is necessary to be mindful, to remain sensitive to the possibility of failure, or to have a constant sense of unease. This can be done by first trying to think of – or even to make a list of – undesirable situations and imagine how they may occur. And then to think of ways in which they can either be prevented from happening or be recognised and responded to as they happen. Such thinking is essential for proactive safety management, but it must be done continuously and not just once.

Although Safety–II focuses on things that go right, it is still necessary to keep in mind that things also can go wrong and to 'remain sensitive to the possibility of failure'. But the 'possible failure' is not just that something may fail or malfunction according to a Safety–I view. It is also that the indented outcomes may not obtain, i.e., that we fail to ensure that things go right. Making sure that things go right requires an ongoing concern for whatever is working or succeeding, not only to ensure that it succeeds but also to counteract tendencies to rely on a confirmation bias or to focus on the most optimistic outlook or outcomes.

In order to remain sensitive to the possibility of failure, it is necessary to create and maintain an overall comprehensive view of work – both in the near term and in the long term. This is for all intents and purposes the same as the concept of (collective) mindfulness that is used by the school of thinking called High Reliability Organisations (HRO). An organisation can only remain sensitive to the possibility of failure if everyone pays close attention to how the social and relational infrastructure of the organisation are shaped. Being sensitive to the possibility of failure allows the organisation to anticipate and thereby prevent the compounding of small problems or failures by pointing to small adjustments that can dampen potentially harmful combinations of performance variability. Many adverse outcomes stem from the opportunistic aggregation of short cuts in combination with inadequate process supervision or hazard identification. Being sensitive to what happens, to the ways in which work can succeed as well as the ways in which it can fail, is therefore important for the practice of Safety–II.

The Cost of Safety, the Gain from Safety

Spending significant amounts of time to learn, think, and communicate is usually seen as a cost. Indeed, money spent on safety is considered a cost or at least a lost opportunity for investment in something else such as productivity. According to Safety–I, an investment in safety is an investment in avoiding that something will happen, not unlike insurance. We know what the cost is, since it is real and regular. But we do not know what the gain or benefit is, since it is both uncertain and unknown in size. In the risk business, the common adage is 'if you think safety is expensive, try an accident'. And if we calculate the cost of a major accident, such as Deepwater Horizon or Fukushima Daiichi, almost any investment in safety would have been cost-effective. However, since we cannot prove that the safety precautions actually worked (only politicians seem to be able to do that with impunity), and since we cannot say *when* an accident is likely to happen, calculations are biased in favour of reducing the investment. (This is something that is typically seen in hard times. It defies logic because the probability of an accident is not reduced when the economic conditions get worse. On the contrary, it is actually increased, since people are pressed harder, are less likely to report incidents, are more likely to take changes, etc. The whole behaviour is therefore counterintuitive.)

In Safety–I, investing in safety is seen as a cost and as non-productive. The investment is made to *prevent* something from happening, rather than to *make* something happen or *improve* the way in which something happens. The decision alternatives are therefore to make the investment or not to make it. The outcome of interest is whether the organisation will be free of accidents for a given time period, say *m* months or *n* years. The two possible outcomes are therefore whether an accident happens during the time period or whether there is no accident. (For the sake of simplicity, the example only considers the occurrence of a single accident.) This altogether produces a matrix with four cells, as shown in Table 8.2. If it is decided to make the investment and there is an accident, then it is seen as a justified investment (on the assumption that the magnitude of negative outcomes will be reduced). If it is decided to make the investment and no accident happens, then it is seen as an unnecessary cost – at least from a business perspective. (One reason is that it cannot logically be argued that the absence of accidents was due to the safety investment.) If the investment is not made and there is an

Table 8.2   The decision matrix for Safety–I

		Anticipated events and outcome values	
		Accidents	**No Accidents**
Decision alternative	Investing in risk reduction	Justified investment	Unnecessary cost
	Not investing in risk reduction	Bad judgement, bad luck	'Justified' saving
Estimated likelihood of an accidents		p	1-p

accident, it is seen as a case of bad judgement or bad luck. And finally, if no investments are made and no accident happens, then it is seen as a justified saving.

If we can estimate the values of the probability of an accident (p), the cost of a safety investment (*I*) and the cost incurred by an accident (*A*), then the decision matrix can be used to rank order the four alternatives.

- If a safety investment is made and if an accident does not occur, the investment is lost; the value of the alternative is therefore -*I*.
- If a safety investment is made, and if an accident occurs, then the investment is seen as justified. The value of the alternative is then -*A* * p + *I*.
- If a safety investment is not made, and if an accident occurs, then the situation is a case of bad luck (or a bad judgement). The value of the outcome is -*A* * p.
- Finally, if a safety investment is not made and if an accident does not occur, then the value is 0.

Given these values, the option of not investing in safety dominates the option of investing in safety.

In Safety–II, an investment in safety is seen as an investment in productivity, because the definition – and purpose – of Safety–II is to ensure that as many things as possible go right. Thus if an investment is made and there is no accident, everyday performance will still be improved. If there is an accident, the investment will again be seen as justified, because of both the

Table 8.3   The decision matrix for Safety–II

		Anticipated events and outcome values	
		No accidents	**Accidents**
Decision alternative	Investing in improved performance	Better everyday performance	Justified investment
	Not investing in improved performance	Acceptable performance, no gain	Bad judgement, bad luck
Estimated likelihood of accidents		1-p	p

improved performance and the reduced consequences. If no investments are made and there is no accident, performance may remain acceptable but will not have improved. And finally, if no investment is made and an accident occurs, it is seen as bad judgement. This is shown in Table 8.3 on the previous page.

If we repeat the above argumentation, a safety investment is no longer a cost but an investment in productivity. We shall modestly assume that the ratio of the investment to productivity improvement is 1:1. If the probability of an accident again is (p) and the cost incurred by an accident is (A), then the following rank order will be the result.

- If a safety investment is made, and if an accident does not occur, then the investment will still have increased productivity; the value of the alternative will be I.
- If a safety investment is made, and if an accident occurs, then the increased productivity will be reduced by the expected loss from the accident. The value of the alternative is I-A * p.
- If a safety investment is not made, and if an accident occurs, then the situation is a case of bad luck (or a bad judgement). The value of the outcome is -A * p.
- Finally, if a safety investment is not made and if an accident does not occur, then the value of the alternative is 0.

Given these values, the option of investing in Safety–II dominates the option of not investing in Safety–II, whereas the same was not the case for Safety–I. The conclusion is therefore that it is better to invest in Safety–II than not to do so.

Comments on Chapter 8

The degree of control that a person has over a situation can vary. This can be characterised by distinguishing between four different control modes called strategic, tactical, opportunistic, and scrambled. The control modes differ in terms of the number of goals that can be considered, the time available, the criteria for evaluating outcomes, and the thoroughness of how the next action is selected. A detailed description can be found in Hollnagel, E.

and Woods, D.D. (2005), *Joint Cognitive Systems: Foundations of Cognitive Systems Engineering*, Oxford: Taylor & Francis.

Although not referred to earlier in this book, Safety–II and resilience engineering represent the same view on safety. For resilience engineering, failure is the result of the adaptations necessary to cope with the complexity of the real world, rather than a breakdown or malfunction. The emphasis must therefore be on how individuals and organisations continually adjust what they do to the current conditions because resources and time are finite. Rather than list the several books written on resilience engineering, readers are recommended to visit https://www.ashgate.com/default.aspx?page=1779.

Complex Adaptive Systems (CAS) is frequently used as a label for the tightly coupled, non-linear systems that Charles Perrow described in his book, *Normal Accidents*. While it is indisputable that many of the systems we have built truly are complex, multifaceted, self-organising and adaptive, the 'CAS' term is frequently used as a convenient label that apparently solves a multitude of scientific problems but which in reality is just another efficiency–thoroughness trade-off. This is clearly recognised by the following quotation:

> More generally, the term 'complexity' is 'present' and doing metaphorical, theoretical and empirical work within many social and intellectual discourses and practices besides 'science'

It can be found in Urry, J. (2005), The complexity turn, *Theory, Culture & Society*, 22(5), 1–14.

Another way of learning about how everyday work takes place is by means of 'extended leave'. This is a procedure where a person assumes the duties of another person, often in a different department. The practice has been described in Fukui, H. and Sugiman, T. (2009), Organisational learning for nurturing safety culture in a nuclear power plant, in E. Hollnagel, (ed.), *Safer Complex Industrial Environments*, Boca Raton, FL: CRC Press. The primary purpose of this practice is, of course, to disseminate and improve safety culture and not to get information for a Safety–II study.

What-You-Look-For-Is-What-You-Find (WYLFIWYF) has been used to name the well-known fact that the data we find depends on what we look for, which in turn depends on how we think of and describe the reality we work with. An example of that is described in Lundberg, J., Rollenhagen, C. and Hollnagel, E. (2009), What-You-Look-For-Is-What-You-Find – The consequences of underlying accident models in eight accident investigation manuals, *Safety Science*, 47(10), 1297–311.

Practical guidance on how to apply Appreciate Inquiry can be found in several books and other publications, such as Cooperrider, D.L., Whitney, D. and Stavros, J.M. (2008), *Essentials of Appreciative Inquiry*, Brunswick, OH: Crown Custom Publishing, Inc. A description of co-operative inquiry, or collaborative inquiry, can be found in Heron, J. (1996), *Cooperative Inquiry: Research into the Human Condition*, London: Sage. (For the cognoscenti, appreciative inquiry can be seen as representing positive psychology, while co-operative inquiry represents action research.) Finally, exnovation has been proposed as a way to make the sources and dynamics of existing practice visible and tangible. A recent, comprehensive description can be found in Iedema, R., Mesman, J. and Carroll, K. (2013), *Visualising Health Care Practice Improvement: Innovation from Within*, Milton Keynes: Radcliffe Publishing Ltd.

Chapter 9
Final Thoughts

The Numerator and the Denominator

In a Safety–I view, the focus is on adverse events. This can be as either the absolute number of adverse events or as the relative number. The focus on the absolute number is a corollary of the myth that all accidents are preventable, usually expressed as the goal or ideal of 'zero accidents', 'zero work site injuries', or even 'zero car accidents'. (To be fair, this is presented as a 'virtually zero car accidents'.) The idea of zero accidents can be found in the mission statements of many large companies, as well as a service offered by numerous consultants. The focus on the relative number corresponds to a more probabilistic view, for instance that the probability of an accident must be below a certain value, or that there should be 'freedom from unacceptable risks'.

In the first case, the pursuit of absolute safety, the concern is for the value of the numerator N, where N represents the number of a certain type of events, for instance accidents, incidents, injuries, lost time injury, unplanned outages; rejection rate, etc. The goal is to reduce N to as small a value as is practicable, preferably zero if that is possible. In the pursuit of absolute safety, there is no concern for the number of complementary events, i.e., the number of cases where N does not occur. For example, if N is the number of times per year a train passes a red signal (SPAD), then the number of complementary events is the number of times during the same time period that a train stops at a red signal. Having $N=0$ as the ideal in practice means that safety is defined as the 'freedom from risk' rather than as the 'freedom from unacceptable (or unaffordable) risk'. The difference is by no means unimportant.

In the second case, the pursuit of relative safety, the concern is for the ratio $\frac{N}{M}$, where M is the number of complementary events. The goal is to make the ratio as small as possible, but recognising that not all accidents are preventable. In this case it becomes important to know something about the numerator, N, and the denominator, M. To look at the denominator can be a first step towards a Safety–II perspective, since it naturally raises the question of what characterises these M events. In other words, it becomes important to understand not only the adverse outcomes that are represented by N but also the complementary outcomes, the activities that went well and that are represented by M.

Focusing on the ratio also shows that there are two ways of making the ratio smaller. The traditional approach, corresponding to Safety–I, is to try to reduce the value of N by trying to eliminate possible adverse outcomes. The futility of relying on this as the only means to improve safety is widely recognised, although it does not always show itself in safety management practices. Another approach, corresponding to Safety–II, is to try to increase the value of M. Trying to increase the number of things that go right, as an absolute or as a relative number, will not only make the ratio $\frac{N}{M}$ smaller, but will also increase the value of M. This is tantamount to making the organisation better at doing what it is intended to do, hence to improve its productivity as well as its safety. It is another way of supporting the argument in Chapter 8 that using resources to improve Safety–II is an investment rather than a cost.

Is the Devil in the Details?

In Safety–I, the concern in accident analyses and risk assessments is to be so thorough and exhaustive that nothing important is overlooked. If something is missed in either kind of analysis, and if it later turns out to play a role in another case, then blame will be shared widely and generously. (The frequently heard defence is the litany that it was never imagined that something like that could happen.) An honest answer to the question of whether all possible risks and hazards have been found, and whether all possible – or even all major – ways in which something can go wrong have been considered, must be in the negative. Murphy's

Law makes it clear that 'anything that can go wrong, will go wrong' – and anything, in practice, means everything. But it is impossible to think of everything that can go wrong for very practical reasons such as limitations in terms of time and money, lack of imagination, stereotypes (default assumptions), strategic interests, expediency, etc.

If, on the other hand, we ask whether we have found all the ways in which something can go right, then the honest answer must again be that we have not, and for much the same reasons. But in this case the lack of completeness is less worrying. The practical limitations are still the same, but if something goes right in an unexpected way it is rarely seen as a problem but rather as an opportunity for improvement and for learning.

It is a general saying that the devil is in the details. This means that the details are important and that the real problems or difficulties are hidden there. That is why we in Safety–I carefully dissect accidents to find the real reasons for them and painstakingly try to explore every logical consequence of how an event can develop. Yet if we adopt a Safety–II perspective, the devil is no longer in the detail but in the whole. The devil is, in fact, not where we look, but where we fail to look. And we commonly fail to look at the whole and to consider the ways in which systems, organisations and socio-technical habitats usually work. But the whole cannot be understood by the usual approach, by decomposing it into details. The whole can only be understood by finding ways to describe it qua itself.

One consequence of looking for the devil in the details is the need for large amounts of data about accidents and incidents, typically organised into large databases. One practical consequence of using Safety–II as a basis for safety management is therefore that it reduces the need for such large databases. When we look at things that go wrong and collect incident reports, each incident is treated as a single and special case and must be recorded accordingly. Databases for incidents, for instance in aviation or in health care, are therefore huge and easily run into several hundred thousand reports. As an illustration of that, there were 41,501 reported patient safety incidents in Denmark in 2012; the estimated number of reportable incidents was more than one order of magnitude larger, namely about 650,000. Reporting, let alone analysing, even 41,500 incidents is not practically feasible. Fortunately, it is unnecessary to accumulate hundreds of thousands of individual cases when we look for how things go right. Instead it is sufficient to develop a description of the daily activity and its expected variability, which means one generic case instead of many specific ones.

A second practical consequence is that there is less need for reporting on failures and accidents, hence less need to ensure that such reports are given. The need to report on accidents and incidents has turned out to be a problem for Safety–I, because of the perceived dominance of the human factor. As humans increasingly came to be seen as the root cause of accidents – typical figures being in the order of 80–90 per cent – safety came to depend on people reporting what went wrong. Since that might easily end by blaming people for what they had done, interpreting any difference between Work-As-Imagined and Work-As-Done as a violation or a 'human error', there predictably developed a reluctance to divulge such information. The solution was to protect people against possible blame by introducing what is called a 'just culture', already mentioned in Chapter 4 This development has been described in the following excerpt from the *Strategic Plan For Safety* developed by the UK National Air Traffic Services in 2004.

> NATS' safety performance is measured through a comprehensive incident reporting and investigation process. Safety incidents arising from ATC operations are reported through the Mandatory Occurrence Report (MOR) scheme operated by the Civil Aviation Authority's Safety Regulation Group (SRG), and investigated and assessed by both NATS and, if necessary, by the SRG. NATS is committed to maintaining a 'just' reporting culture to ensure that all safety related incidents continue to be reported and investigated.

The problem for a Safety–I perspective is that we live in a complicated, intricate world, where work takes place in conditions of multiple interacting technical, financial, cultural and political constraints. Doing things perfectly under such conditions is hardly a feasible option. But it is difficult for safety management to adopt a view that involves complicated trade-offs, since that does not blend well with the ideal of a well thought-through endeavour, driven by scientific knowledge and practices, and conducted by rational people. The safety myths described in this book all derive from that ideal. As myths, they are counter-productive because they lead to unrealistic safety management attitudes, policies and targets. They also lead to a preoccupation with details and a disregard for the whole. In order to have any chance of successfully operating increasingly complicated socio-technical systems and managing intractable socio-technical habitats, we need to abandon the myths and the idealised approach to safety that they imply.

Second Stories vs The Other Story

When an accident has happened, when something has gone wrong and there is an undesired outcome, the initial reaction is to try to find out what happened and why. This traditionally means trying to find – or produce – an acceptable explanation for what happened, specifically to find a cause or set of causes. Since the search for such causes usually takes place with some urgency, there is a nearly unavoidable tendency to focus on what stands out in the situation, on the factors or features that are easy to see. This practice was in 2002 named the search for 'first stories' in a seminal paper by David Woods and Richard Cook called 'Nine steps to move forward from error', already mentioned in Chapter 6. According to these authors, first stories represent a reaction to failure that narrowly looks for proximal factors – typically the so-called 'human error'. This is consistent with the *causality credo*, according to which the cause of an adverse outcome must be a failure or malfunction of some kind. Since all systems are socio-technical systems, there will always be some humans who, directly or indirectly, are involved. And since humans are 'known' to be prone to fail – to be fallible machines – it makes good sense to accept that as a cause, at least according to this way of thinking. Because human performance is always variable – and, indeed, always must be variable, as is argued throughout this book – the search for a 'human error' is bound to meet with success. The search is therefore rewarded, in the language of learning theory, and the response is reinforced, which makes it more likely to be used the next time there is a similar issue. This quickly leads to a self-sustaining mode of responding that soon becomes a habit. Woods and Cook described it thus:

> When an issue breaks with safety at the centre, it has been and will be told as a 'first story'. First stories, biased by knowledge of outcome, are overly simplified accounts of the apparent 'cause' of the undesired outcome. The hindsight bias narrows and distorts our view of practice after-the-fact.

In order to combat this bias, it is necessary to look for 'second stories', i.e., for other possible explanations of what happened. This can be facilitated in a number of ways, such as acknowledging that there is a hindsight bias, understanding how work is shaped by conditions, and searching for underlying patterns. It can also be facilitated by using a breadth-first approach rather than a depth-first approach (Chapter 8).

While going beyond the first story to look for second stories – and beyond the second story to look for third stories and so on – is a step forward, it nevertheless remains limited by the very way in which it begins. The search is for a cause, hence for an explanation of why something went wrong and why there was an undesired outcome. Looking for second stories is therefore still a part of a Safety–I perspective, although it is a less rigid and more constructive way of going about it. Safety–II will, however, start from a different position because it tries to understand how everyday work succeeds. It is no longer sufficient to look at what has gone wrong and try to find acceptable explanations for that. It is instead necessary to look at everyday performance as a whole, to understand how it succeeds, and to use that as the basis for understanding the cases where it did not. The full consequence of Safety–II is to look for the 'other story' rather than for the 'second story', where the 'other story' is the understanding of how work succeeds. This will of course include a search for 'second stories', since that is a way to realise that there always is more than meets the eye, and a way to be mindful of one's own thinking.

What About the Name?

The first public description of the S–I, S–II distinction coincided with the launch of the website for the Resilient Health Care Net (www.resilienthealthcare.net) on 18 August 2011. It was followed soon after by an article in the programme note for *Sikkerhetsdagene*, a safety conference in Trondheim, Norway, on 10–11 October.

The idea of contrasting two approaches to safety was itself inspired by a similar debate that took place within the field of Human Reliability Assessment (HRA). In 1990 the Human Reliability Analysis (HRA) community was seriously shaken by a concise exposure of the lack of substance in the commonly used HRA approaches (Dougherty, E.M. Jr. (1990), Human Reliability Analysis – where shouldst thou turn? *Reliability Engineering and System Safety*, 29, 283–99). The article made it clear that HRA needed a change, and emphasised that by making a distinction between the current approach, called first-generation HRA, and the needed replacement, called second-generation HRA.

Another well-known use of this rhetorical device is the juxtaposition of Theory X and Theory Y in Douglas McGregor's 1960 book *The Human Side of Enterprise*. The juxtaposition was used to encapsulate a fundamental distinction between two different management styles (authoritarian and participative, respectively), which turned out to be very influential. And there are, of course, even more famous examples, such as Galileo's *Dialogue Concerning the Two Chief World Systems* and the philosophical dialogue in the works of Plato.

Will There Be A Safety–III?

Since Safety–II represents a logical extension of Safety–I, it may well be asked whether there will not someday be a Safety–III. In order to answer that, it is necessary to keep in mind that Safety–I and Safety–II differ in their focus and therefore ultimately in their ontology. The focus of Safety–I is on things that go wrong, and the corresponding efforts are to reduce the number of things that go wrong. The focus of Safety–II is on things that go right, and the corresponding efforts are to increase the number of things that go right.

Safety–II thus represents both a different focus and a different way of looking at what happens and how it happens. Doing this will, of course, require practices that are different from those that are commonly used today. But a number of these practices already exist, either in principle or in practice, as described in Chapter 8, and can easily be taken into use. It will, of course, also be necessary to develop new methods and techniques that enable us to deal more effectively with what goes right, which are able in particular to describe, analyse, and represent the ubiquitous performance adjustments.

If the way ahead is a combination of the existing practices of Safety–I with the complementary – and to some extent novel – practices of Safety–II, then where does that leave a possible Safety–III? It has been suggested that Safety–III 'simply' stands for the combination of existing and novel practices. But the combination of practices is not against the idea of Safety–II, which is intended as a complement to Safety–I rather than a replacement for it (cf., Figure 8.1). Neither does the suggestion of a possible 'Safety–III' offer a new understanding of safety, a new ontology, in the way that Safety–II does, and it may therefore not be necessary in the same way. It can, of course, not be ruled out that in some years' time there may come a proposal for understanding safety with a definition of its own that is different from both Safety–I and Safety–II. It may also happen that the very concept of safety is gradually dissolved, at least in the way that it is used currently, as something distinctively different from, e.g., quality, productivity, efficiency, etc. If that happens – and several signs seem to indicate that it will – then the result will not be a Safety–III but rather a whole new concept or synthesis (see below). So while Safety–II by no means should be seen as the end of the road in the efforts to ensure that socio-technical habitats function as we need them to, it may well be the end of the road of safety as a concept in its own right.

From Safety Analysis to Safety Synthesis

The predominant approach to safety relies on analysis, as in accident analysis or risk analysis. This is in good accordance with the Western tradition of science that for better or worse has brought us to where we are now. In relation to safety this tradition is found, for instance, in the *causality credo*, and in the ontology and aetiology of Safety–I. But an analytical approach is neither inevitable nor the only choice, as the presentation of Safety–II has argued. Indeed, the basic idea of Safety–II is to build or construct something, rather than to take something apart. Safety–II thereby implies that there is an alternative to safety analysis, namely safety synthesis.

The meaning of the noun *synthesis* (from the ancient Greek σύνθεσις) is a combination of two or more entities that together results in something new; alternatively, it may mean the activity of creating something out of something else that already exists. The meaning of 'safety synthesis' is, therefore, the system quality that ensures that things go right, or the system's ability to succeed under varying conditions, so that the number of intended and acceptable outcomes is as high as possible. This is, however, not a natural and stable condition, but an artificial and potentially unstable one. Safety is, to paraphrase Weick's terminology, a 'dynamic event', hence something that must be created constantly and continuously. The basis is that which makes up everyday work and everyday existence. The synthesis of this, the bringing together of what individuals and organisations do on all levels and over time, is what creates safety – at least until a better term has been adopted. This synthesis has two different forms, a synthesis across organisational levels and a synthesis over time.

The synthesis across levels is relatively easy to explain. It means that one must understand the dependencies or couplings between everything that happens in an organisation or in carrying out an activity, no matter which levels of the organisation or types of work are involved. And it is, of course, necessary that the people who do the work understand that themselves, i.e., that the synthesis is part of what everyone does, or at least that it is recognised by them.

The synthesis over time is more difficult to explain, but no less essential. In many kinds of activities – probably in all – synchronisation is important. This certainly goes for industries where safety (whether as Safety–I or as Safety–II) is a concern; it goes for services; for communication; for production – not least if it is lean; and so on. Synchronisation is achieved by organising the various productive processes to avoid delays (outputs arriving too early or too late), to ensure a better use of the resources (for instance, doing things in parallel so that the same preconditions do not have to be established twice), to coordinate transportation of matter and energy between processes and sites, and so on. But synchronisation is not the same as synthesis. A synthesis is first achieved when we understand how things really fit together, when we understand the variability of everyday performance (approximate adjustments), and how this variability may wax and wane, leading to outcomes that are sometimes detrimental and sometimes beneficial. A temporal synthesis cannot be achieved by exploring pairwise combinations of functions, even if it is done exhaustively. Neither can it be achieved by a bottom-up approach. Instead it requires a genuine top-down perspective in the sense of being able to see (by serendipity rather than by combinatorics) when something new and useful has been achieved. Safety synthesis is the constant creation and maintenance of conditions that allow work to succeed on the basis of all criteria taken together.

Glossary

(Approximate) Adjustments When working conditions are underspecified or when time or resources are limited, it is necessary to adjust performance to match the conditions. This is a main reason for performance variability. But the very conditions that make performance adjustments necessary also mean that the adjustments will be approximate rather than perfect. The approximations are, however, under most conditions good enough to ensure successful performance.

Aetiology Aetiology is the study of causation and of why things occur, or even the study of the reasons or causes behind what happens. In relation to safety it is the study of the (assumed) reasons for or causes of the observable phenomena. The aetiology describes the 'mechanisms' that produce the observable phenomena, the phenomenology.

Bimodality Technological components and systems are built to function in a bimodal manner. Strictly speaking, this means that for every element e of a system, the element being anything from a component to the system itself, the element will either function or it will not. In the latter case the element is said to have failed. The bimodality principle does, however, not apply to humans and organisations. Humans and organisations are instead multi-modal, in the sense that their performance is variable – sometimes better and sometimes worse but practically never failing completely. A human 'component' cannot stop functioning and be replaced in the same way a technological component can.

Efficiency–thoroughness trade-off The efficiency–thoroughness trade-off (ETTO) describes the fact that people (and organisations) as part of their activities practically always makes a trade-off between the resources (time and effort) they spend on preparing an activity and the resources (time, effort and materials) they spend on doing it, usually to the disadvantage of the former.

Emergence   In a growing number of cases it is difficult or impossible to explain what happens as a result of known processes or developments. The outcomes are therefore said to be emergent rather than resultant. Emergent outcomes are not additive, not decomposable into 'components' and, consequently, not predictable from knowledge about such 'components'.

Intractable systems   Systems are called intractable if it is difficult or impossible to follow and understand how they function. This typically means that the performance is irregular, that descriptions are complicated in terms of parts and relationships, and that it is difficult to understand the details of how the system works, not least because it changes faster than a description can be completed. Intractable systems are also underspecified, meaning that it is impossible to provide a complete specification of how work should be carried out for a sufficiently large set of situations.

Ontology   The ontology describes that 'which is'. In this book, ontology is the study of what the foundation of safety *is*, rather than how it *manifests* itself (the phenomenology) or how these manifestations *occur* (the aetiology).

Performance variability   The contemporary approach to safety (Safety–II) is based on the principle of equivalence of successes and failures and the principle of approximate adjustments. Performance is therefore in practice always variable. The performance variability may propagate from one activity or function to others and thereby lead to non-linear or emergent effects.

Phenomenology   The phenomenology refers to the observable characteristics or the manifestations of something: in this book, it is that which makes us say that something is safe – or unsafe.

Resilience   A system is said to be resilient if it can adjust its functioning prior to, during, or following changes and disturbances, and thereby sustain required operations under both expected and unexpected conditions.

Resilience engineering   The scientific discipline that focuses on developing the principles and practices that are necessary to enable systems to function in a resilient manner.

Safety–I   Safety is the condition where the number of adverse outcomes (accidents/incidents/near misses) is as low as possible. Safety–I is achieved by trying to make sure that things do not go wrong, either by eliminating the causes of malfunctions and hazards, or by containing their effects.

Safety–II   Safety is a condition where the number of successful outcomes is as high as possible. It is the ability to succeed under varying conditions. Safety–II is achieved by trying to make sure that things go right, rather than by preventing them from going wrong.

Socio-technical system   In a socio-technical system, the conditions for successful organisational performance – and, conversely, also for unsuccessful performance – are created by the interaction between social and technical factors rather than by either factor alone. This interaction comprises both linear (or trivial) 'cause and effect' relationships and 'non-linear' (or non-trivial) emergent relationships.

Socio-technical habitat   A socio-technical habitat is the set of mutually dependent socio-technical systems that is necessary to sustain a range of individual and collective human activities. A workplace can be described as a socio-technical system when it is considered by itself. But its sustained functioning always depends on inputs and support provided by other socio-technical systems, for instance in relation to transportation, distribution, communication and control, etc. Combined socio-technical systems constitute a socio-technical habitat or a macro-social system.

Tractable systems   Systems are called tractable if it is possible to follow and understand how they function. This typically means that the performance is highly regular, that descriptions are relatively simple in terms of parts and relationships, and that it is easy to understand the details of how the system works, not least because it is stable.

Index

Printed in the United States
by Baker & Taylor Publisher Services